# Payment schemes for forest ecosystem services in China: policy, practices and performance

# Payment schemes for forest ecosystem services in China: policy, practices and performance

Dan Liang

*Environmental Policy Series – Volume 5*

*Wageningen Academic*
P u b l i s h e r s

Wageningen Academic Publishers
P.O. Box 220
6700 AE Wageningen
The Netherlands
www.WageningenAcademic.com
copyright@WageningenAcademic.com

ISBN: 978-90-8686-199-6
e-ISBN: 978-90-8686-754-7
DOI: 10.3920/978-90-8686-754-7

First published, 2012

© Wageningen Academic Publishers
The Netherlands, 2012

# Preface

This book attempts to provide an evaluation on China's payment policies for forest ecosystem services. At the end of last century, the Chinese government launched an array of ecological conservation projects, which reshaped China's forest governance. Since I started to work in China National Forestry Economics and Development Research Center (FEDRC) in 2003, I have been involved into several projects, which aimed to evaluate the impacts of these ecological conservation programs on local economy and society. This research gave me opportunity to have a first look into the implementation of these ecological orientated programs and their consequences. However, this research relied on statistical data, but paid little attention to the policy process in local context. At the same time, I also had several trips to forested regions each year and my conversations with local farmers showed me a far more complex picture than what the statistical data described. When I began my PhD study at the Environmental Policy group (ENP), Department of Social Sciences, Wageningen University in 2007, I was fascinated by the insights of Ecological Modernization Theory and its sociological perspective. The PhD research gave me an opportunity to reflect on my previous work and to develop a new angle to examine payment policies for forest ecosystem services in China.

This research has been implemented in the framework of the "Governmental Environmental Auditing on Ecological Programs in China" project, a collaborative program funded jointly by the Royal Netherlands Academy of Arts and Sciences (KNAW) in the Netherlands and the Chinese Academy of Sciences (CAS) in China. I am grateful to the financial support from both sides.

This thesis could only be accomplished with invaluable help from many people. I am deeply indebted to Professor Arthur P.J. Mol and Professor Jan van Tatenhove, my supervisors. I would like to express my sincere thanks for their valuable advice, guidance, encouragement and patience throughout my research. Their expertise and support is crucial for my PhD study. I also highly appreciate Prof. Yonglong Lu, my co-supervisor in the Research Center for Eco-Environmental Sciences (RCEES), CAS. With his trust and recommendation, I started to apply for this PhD candidate position and with his guidance and support I could smoothly finish my field research in China.

I enjoyed studying and working at the Environmental Policy group with my colleagues. They offered such a stimulating and friendly working environment that I could make constant progress in my PhD study. I would especially like to thank Corry Rothuizen, who made my life and study at Wageningen much easier. I owe many thanks to Zhang Lei for her support during my research and the whole project. ENP is not only a place for open and free academic research, but also a warm home for its international community. I appreciate the encouragement and help from my colleagues: Marjanneke Vijge, Elizabeth Sargant, Dorien Korbee, Bettina Bluemling, Loes Maas, Natapol Thongplew, Ali Haider, Carolina Maciel, Jennifer Lenhart, Eria Bieleveldat-Carballo Cardenas, Ching Kim, Kanang Kantamaturapoj, Sammy Letema, Leah Ombis, Tung Son Than, Harry Barnes Dabban, and Alexey Pristupa. I would especially like to thank Marjanneke for helping me with the Dutch translation included in this thesis.

When I stayed at Wageningen, I met many Chinese friends. Their support and encouragement made my life abroad less tough and more colorful than I could image. My thanks go to Zhong

Lijin, Han Jingyi, Zhang Yuan, Li Feng, Feng Yan, Jin Shuqin, Liu Wenling, Lu Jing, Wu Yan, Li Yuan, Li Hui, Wu Jing, Song Yanru, Zhang Yunmeng, Tian Lijin, Li Zhaoying, Zhang Lei, Tu Qin, Qu Wei, Bin Xiaoyun, Li Jia, Zheng Chaohui, Guan Ye, Liu Xiao, and Liu Wei.

My research in China also benefited from friends and fellow students in RCEES. Seminars and discussions with them inspired me to improve my study. Special thanks go to Shi Yajuan, He Guizhen, Wang Tieyu, Luo Wei, and Yuan Jingjing.

I owe many thanks to my colleagues at the State Forestry Administration. Zhang Lei, Dai Guangcui, and Tang Xiaowen helped me with getting local contacts and arranging field work. Wang Huanliang, and Wang Yuehua gave me valuable comments and suggestions on my research. Xie Chen, Zhang Sheng, Zhang Zhitao, Zhao Jincheng, Gu Zhenbin and Li Jie shared with me their research experiences and insights on forest management and protection in China.

Finally, I would like to dedicate this book to my parents, to thank their unreserved love and support over these years.

Beijing, February 2012

# Table of contents

## Chapter 6.
## Payment schemes in the transition of collective forest tenure in Liaoning 129

## Chapter 7.
## Conclusion 159

# Abbreviations

| | |
|---|---|
| ACF | Advocacy Coalition Framework |
| APEC | Asia-Pacific Economic Cooperation |
| ARPC | Autonomous Regional People's Congress |
| CBA | cost benefit analysis |
| CCCPC | Central Committee of Chinese Communist Party |
| CCFGP | Conversion of Cropland into Forest and Grassland Program |
| CEA | cost-effectiveness analysis |
| CFB | County Forestry Bureau |
| CFTR | collective forest tenure reform |
| CPC | Communist Party of China |
| EEA | European Environmental Agency |
| EM | ecological modernization |
| EMT | Ecological Modernization Theory |
| FAO | Food and Agricultural Organization of the United Nations |
| FEBCFP | Forest Ecological Benefit Compensation Fund Program |
| FES | forest ecosystem services |
| FHPDP | High-yielding Plantation Development Program |
| FYP | five year plan |
| GDP | gross domestic product |
| GONGO | government-organized NGOs |
| IAD | institutional analysis and development |
| MEP | Ministry of Environmental Protection |
| MFB | Municipal Forestry Bureau |
| MoF | Ministry of Finance |
| NEPO | National Environmental Protection Office |
| NFPP | Natural Forest Protection Program |
| NPC | the National People's Congress |
| PBF | public benefit forest |
| PEPB | Provincial Environmental Protection Bureau |
| PES | payment for environmental services |
| PFD | Provincial Forestry Department |
| PPC | Provincial People's Congress |
| PPPCC | Provincial People's Political Consultative Conference |
| PRC | People's Republic of China |
| RMB | Chinese currency |
| SBP | Shelterbelt Program in Three North area and along Yantze River |
| SC | State Council of People's Republic of China |
| SCPVBT | Sandification Control Program in the Vicinity of Beijing and Tianjin |
| SDRC | State Development and Reform Commission |
| SFA | state forestry administration |

| SOFE | state-owned forestry enterprise |
| SOFF | state-owned forest farm |
| TFS | township forestry station |
| USSR | Union of Soviet Socialist Republics |
| VC | Village Committee |
| WCNRDP | Wildlife Conservation and Nature Reserve Development Program |
| WTO | World Trade Organization |

# Chapter 1.
# Ecological crisis, forest protection and payment schemes

## 1.1 Introduction

China has a complex and diverse natural and geographical environment. Its forest resources, with numerous species and vegetation types, provide various products and ecological services. However, the forest resources in China are under pressure from its fast-growing economy and rapid industrialization. There is a historical transition in China's forest sector from traditional forest management to forest governance. The emergence of payment schemes for forest system services around the country signifies this significant change.

### 1.1.1 China's forest resources

China's forests cover 195.45 million hectares, with a forest volume 13,721 million cubic meters (State Forestry Administration, 2009c). Although China's forest area accounts for 5.12% of the world's total forest area, and ranks fifth behind Russia, Brazil, Canada and USA, its forest area per capita is quite low: only 25% of the average per capita forest area of the world (FAO, 2011). The demand of society on timber production imposes still a huge pressure on forest resources in the country.

Although the pressure on forest resources was imposed by its fast industrialization, from the late 1970s to 2008, China's forest still extended almost in all provinces, due to the development of planation and the protection on natural forest (see Figure 1.1). The natural forest area in China is 119.69 million hectares with a stock volume of 11,402 million cubic meters (State Forestry Administration, 2009c). Natural forest was the main industrial base for timber production before 1997, but after the start of the Natural Forest Protection Program (NFPP), most natural forest was strictly protected for the purpose of ecological conservation. Compared to the sixth forest inventory (1999-2003), the seventh forest inventory (2004-2008) showed that the natural forest under the program increased both in area (by 26.37%) and in stocking volume (2.23 times) (State Forestry Administration, 2009b). Besides natural forest, China puts a lot of effort to develop plantations to increase its forest resources for both ecological purpose and timber production. In 2009, the plantation area reached 61.69 million hectares with a stock volume of 1,961 million cubic meters, and it had increased by 31.21% since the sixth forest inventory (State Forestry Administration, 2009c). After years of afforestation, China has the largest area of forest plantation in the world (FAO, 2011). The plantations not only produce timber and other forest products, but also provide ecological services for society. Over 40% of the plantation (by area) is timber forest, 31.6% is economic forest, 25.2% is protection forest and the rest are fuel forest and special use forest (State Forestry Administration, 2009b).

Most of China's forest grows in 5 major forest regions, including the northeast and Inner Mongolia forest region, the southwest mountainous forest region, the southeast low mountainous and hilly forest region, the northwest mountainous forest region, and the tropical forest region (State Forestry Administration, 2009b) (see Figure 1.2). These five forest regions cover 40% of the

Forest coverage (1977-1981)                    Forest coverage (2004-2008)

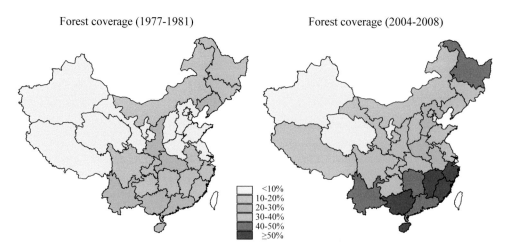

*Figure 1.1. China's forest coverage change from 1977 to 2008. The data based on the second forest inventory from 1977 to 1981 and the latest seventh inventory from 2004 to 2008.*

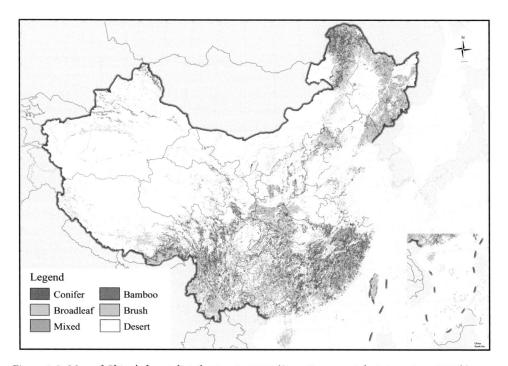

*Figure 1.2. Map of China's forest distribution in 2008 (State Forestry Administration, 2009b).*

total national territory, but they include 70% of the total forest area and 90% of the total forest stocking volume (State Forestry Administration, 2009b). The northeast and Inner Mongolia forest region covers Heilongjiang, Jilin and Liaoning provinces. With humid climate and mild mountains, this region is an ecological barrier for the Songnen Plain, Sanjiang Plain and Hulun Buir Grassland in the northeast China and also an important timber production base. The southwest mountainous forest region includes parts of Yunnan, Sichuan Province and Tibet Autonomous Region. This region is characterized by low latitudes, high elevation, favorable climate and rich and diverse species. The southeast low mountain and hilly forest region has the largest forest area and covers 12 provinces in South China. It lies in the subtropical zone with mild climate and sufficient rainfall, which is suitable for developing timber forest and economic forest. The northwest mountainous forest region covers forests in Xinjiang, Gansu and Shanxi, which are important for the ecologically fragile west region. The tropical forest region locates in some parts of Yunnan, Guangxi, Guangdong, Hainan, and Tibet, where warm climate and plenty rainfall are suitable for the growth of large-diameter trees and diverse species.

### *1.1.2 China's economic development and ecological crisis*

From the 1980s, the world witnessed China's rapid economic growth, benefited from its market-based economic reform. Its GDP increased by more than 9.8% on average each year. At the same time, abundant natural resources, including forest, has been utilized to fuel its economic engine. Figure 1.3 shows the timber production kept increasing and both forest areas and stock volume stayed at a low level during the early period of the reform. As a consequence, ecological degradation occurred along the major watersheds.

Ecological disasters and tremendous damage draw the attention of the Chinese government to environmental issues related to deforestation. Especially the devastating flooding along major rivers in 1998 caught wide public attention to forest degradation. During the 1998 summer, China

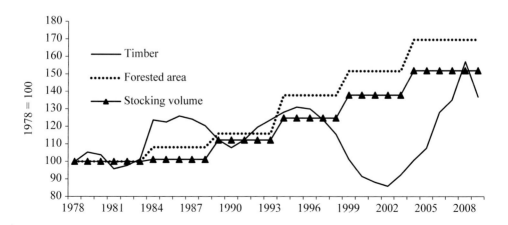

*Figure 1.3. China's forest resource and timber production from 1978 to 2009 (State Forestry Administration, 2009b, 2011).*

suffered the heaviest flooding since 1954, and the flood almost overflowed the entire watershed and 29 provinces were seriously damaged, especially Jiangxi, Hunan, Hubei and Helongjiang (RMW, 2002). The devastating floods killed more than 3,600 people (Lang, 2002), left 14 million people homeless, affected 240 million people, destroyed 5 million houses, damaged 12 million houses, flooded 25 million hectares of farmland, and caused over US$ 20 billion in estimated damages (NCDC, 1998).

After the disaster, many experts analyzed the reason for the flood along the Yangtze River. The examination showed that in spite of abnormal climate, the amount of precipitation over the catchment and the floodwater discharge from the upper basin did not exceed the historical maximum. However, water levels in the middle basin were much higher than the historical maximum (Zong and Chen, 2000). It means that the abnormal climate is not the main factor causing the flood and that the capacity of the river has been decreased. Further research showed that deforestation and reclamation on marginal slope land in the catchment area has induced soil erosion, resulting in a large amount of sediment deposited in reservoirs whose storage capacity is thus reduced (Li, 1999; Zong and Chen, 2000). Soil erosion occurred over 20% of the catchment and 2.4 billion tons of topsoil was lost annually. Soil erosion was the reason that the riverbed of some sections rose many years ago (RMW, 2002). In addition, the canopy and litter of forest can effectively slow down the runoff from heavy precipitation, but the deforestation had damaged that function (Li, 1999).

The public were also aware of the looming ecological crisis from forest degradation. According to the investigation on China's public environmental awareness of forest-related ecological issues by Beijing Lingdian Market and Analysis Company and State Forestry Administration (SFA)[1], more than 90% of the respondents thought that forest degradation and deforestation was the major reason for the 1998 flood along Yangtze River, Songhua River and Neng River (Lingdian Company, 1999).

Since the late 1990s the government has tried to launch a series of ecological restoration projects to protect forests and stop overcutting. However, China's economic engine still had to be fueled by timber processing and timber product exports (mostly wooden furniture, pulp and paper, veneer and wooden board). Due to the reduction on domestic timber production, China's timber processing sector had to rely on timber imports worldwide as raw materials. Since the late 1990s, China's imports of unprocessed wood products have increased substantially. Figure 1.4 illustrates a rapid growth in the volume of timber imports from 1998 to 2009. Both the reduction in domestic supply caused by the ecological restoration projects and the changes in tariff levels after participation in the Asia-Pacific Economic Cooperation (APEC) and accession to the World Trade Organization (WTO)[2] drove the increase of imports from other countries (Zhu *et al.*, 2004).

Among the countries exporting timber to China, Russia, New Zeeland, and Canada were the three largest suppliers (see Figure 1.5). They together supplied more than 25.6 million cubic

---

[1] The investigation has been carried out in February 1999 covering 5 major Chinese cities (Beijing, Shanghai, Guangzhou, Wuhan and Chengdu).

[2] After it joined APEC in 1991 and the WTO in 2001, China loosened controls over most wood-product imports. Tariffs for logs, sawn wood, wastepaper and pulp have been reduced to zero and furniture tariffs, which were 78% in 1992, were reduced to 22% in 1999.

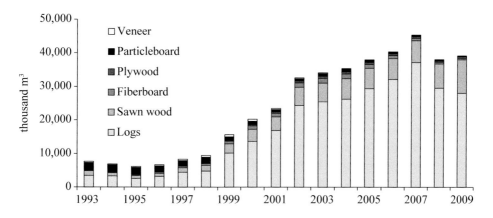

*Figure 1.4. China's timber imports from 1993 to 2009 by product type (State Forestry Administration, 2011).*

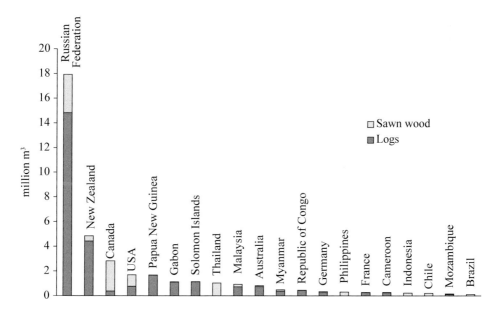

*Figure 1.5. China's imports of major forest products in 2009 by supplying countries (State Forestry Administration, 2010a).*

meters of timber to China in 2009 (State Forestry Administration, 2010a). The United States, Papua New Guinea, Gabon, Solomon Islands, and Thailand make up a second tier of countries that each exported more than 1 million cubic meters of timber to China in 2009 (State Forestry Administration, 2010a). China's ecological transformation in the forest sector and international market forces have led to a surge in the volume of timber, especially logs, imported into China over the last decade. In some of the countries supplying these logs, especially Southeast Asian and African countries, timber production poses enormous threats to the environment (Zhu *et al.*, 2004). Some international environmental NGOs such as Greenpeace and Global Witness claimed that China is one of the major destinations for wood illegally harvested, either at source or somewhere along the supply chain (GW and EIA, 2010; Stark and Cheung, 2006). In the past, China's wood buyers did not show concern for the environmental or legal credentials of their suppliers. Lately, market and regulatory pressure require them to take into account the environmental footprint of timber imported and seek alternative and more ecologically sustainable sources. From 2003 to 2009, more timber imports came from sustainable suppliers such as New Zeeland, Canada and USA. In the total imports, hardwood more and more replaced broadleaf wood, which was usually supplied by tropical countries in Southeast Asia and Africa (State Forestry Administration, 2010a).

### 1.1.3 Historical transitions in forestry

Responding to the public awareness, China's leadership began actions to reverse the forest ecological degradation. The forest department announced that five transitions had to be promoted in China's forest sector due to insufficient ecological services from forest ecosystems: from timber production to ecological development; from felling natural forests to felling plantations; from reclaiming land by deforestation to reforestation; from unpaid to paid utilization of forest ecological benefits; from forest managed by a single sector to full participation by society (Zhou, 2002). From the 1950s to the late 1970s, China's forestry focused on timber production and utilization. All production and consumption in forestry revolved with fulfilling national plans for timber production and providing timber for economic development. From the late 1970s to the 1990s, the demand of ecological services on forest ecosystems became more and more urgent, while timber production was still the major purpose of the forest sector. Ecological conservation programs, such as the Shelterbelt Program in Three-North Area, were launched to protect fragile regions from ecological crisis. But the strong demand of society on timber production has not changed fundamentally. Furthermore, as a way for farmers to improve their livelihood and escape from poverty, timber production was still an important income source for farmers in many regions, especially in the underdeveloped mountainous areas. A consensus on protecting forest for ecological services had not been reached among the public. In the end of the 1990s, the transition in forestry started to materialize and the sector entered into a new phase. First, the implementation of China's strategy for sustainable development, in which forestry is an important component, promoted the idea of environmental protection and ecological conservation among the public. At the same time, with economic growth the public turned their attention to ecological services of forests, rather than just their productive function for timber. Second, the above mentioned devastating flood along the Yangtze River sounded an alarm for the enduring environmental degradation and depletion of forests in the country. Its severe consequences caught the attention for society and the government and facilitated a consensus on an urgent need for forest protection. Third, the increasing fiscal revenue following the 1994 Fiscal Reform strengthened the central government's financial ability to provide pubic goods, including ecological restoration and environmental protection. The central government shouldered the responsibility for ecological restoration around the country.

### *1.1.4 Payments for environmental services around the world*

Payments for environmental services (PES) have been increasingly applied as a mechanism to translate values of ecosystems into incentives for local actors to provide environmental services. The term PES has been used as a broad umbrella for many kinds of instruments for conservation involving cash or in-kind payment for suppliers. Although in theory PES is regarded as a market solution to environmental problems and as an alternative to state and community governance, in practice it needs not only marketed-based mechanisms but also state and community engagement. The reconfiguration of the relationship between the state, market and communities in different contexts results in the diversity of PES. Case studies of PES around the world discovered various instruments for PES in both developed and developing countries (Landell-Mills and Porras, 2002).

By the types of mechanisms for paying environmental serviced, PES can be divided into public payment schemes and market-based schemes. In a market-based PES, the price of ecological services (ES) is often decided in a market in which buyers and suppliers voluntarily exchange ES. Public payment schemes are usually established by the government, which provides payment for suppliers to encourage more provision of ES. In some public payment schemes, the suppliers are required to provide ES rather than voluntarily participate in the schemes, especially in developing countries.

Examples of public payment schemes include governmental purchase in the United States (Forest History Society, 2006), national-scale PES programs in Costa Rica (Rojas and Aylward, 2003) and agri-environmental schemes in Europe (EUROPA, 2008; Forestry Commission, 2008). Market-based payment schemes have more diverse forms, including conservation contract trading by a land trust (LTA, 2006; Merenlender *et al.*, 2004), equity finance by enterprises and NGOs (Sydee and Beder, 2006), eco-service credit exchange (Connora *et al.*, 2008; DECC, 2007; Hamilton *et al.*, 2007), eco-labeling for forest and agricultural products (Gobbi, 2000), and eco-service fees in southern America, south Asia and the USA (IIED, 2004; Landell-Mills and Porras, 2002).

Public fiscal payment schemes and market-based payment schemes have been widely applied to various environmental services from different eco-systems in many countries over different geographical regions. Comparatively, public payments by governments incline to cover large scale areas and are often applied as a policy instrument to the whole nation, while market-based payments run at different levels from global to local. Although some market-based instruments like carbon trade markets and eco-labeling have been adopted at a global level, the majority of them are still quite small in scale, for example, those payment schemes within small watersheds or for private forests with special ecological values. Public payment schemes implemented by governments usually use single payment standards, lack of sufficient evidence for externality, and involve high transaction costs. In effect, their efficiency is compromised, especially for the payment schemes in developing countries. When the payment schemes get scaled up, the government becomes more difficult to know information about the ecological value of the ecosystem and the cost of suppliers. The hidden information raises the cost for the government to design, implement and monitor the payment schemes. However, under market-based mechanisms, buyers for ecological services can obtain the information on the cost of the suppliers by pricing. Therefore, market-based instruments usually can achieve higher efficiency than public payment schemes. But the smooth implementation of market-based instruments requires reasonable arrangements for property rights, mature market for ecological services, transparent information and ancillary laws and regulations. The trends show that in developed countries relatively sophisticated marketed-based instruments have been applied, such as biodiversity credit trade, and carbon credit futures exchange, while developing countries use more simple instruments such as direct negotiations and conservation contracts.

The development of payment schemes for forest ecosystem services shows two trends. First, market-based instruments are increasingly combined with public payment schemes. There is no longer a clear line between these two types. In practice, some public payment schemes start to adopt market-based instruments. For example, the public payment schemes in USA and Australia used auctions as pricing tool to disclose the information on cost and increase the efficiency of the schemes. The other trend is that partnerships between various actors have increasingly emerged in the payment schemes. Governments, NGOs, and enterprises together provide funding for public payment schemes; NGOs such as land trusts commission land with ecological value to governmental agencies after they appropriate the land or make a conservation contract with land owners. The government as a mediator also takes part in negotiations between buyers and suppliers of ecological services and facilitates the implementation of the contract by guaranteeing the payment and monitoring conservation activities. In addition, the government also provides institutional support for the implementation of private trade for ecological services, such as distributing initial credits of ecological services and making fundamental rules for the market.

### *1.1.5 Emerging payment schemes in an era of ecological crisis*

Forest degradation proved showed that the regulation on forest management in the past decades had failed to meet its target to provide a sustainable forest ecosystem. Therefore, during the transitions, forest ecosystem protection has changed from solely relying on regulating and controlling forest logging to integrating a variety of financial instruments to motivate protection and restoration. Figure 1.6 shows the change of central investment in forestry from 1996 to 2009. Since 1998, the central government has increased its investment in forestry and the investment has grown rapidly. The investment at first was supported majorly by treasury bonds, which was out of the central budget and less stable than funds within the central budget. Later, more than half of the investment came from central budget funds. It meant that the source of investment became a stable and fixed item in the central budget. In addition, discounted loans for forestry and infrastructure development capital within the State budget also increased rapidly.

Most of the investment flew into forest protection and ecological conservation programs especially focusing on the Natural Forest Protection Program (NFPP) and the Conversion of Cropland into Forest and Grassland Program (CCFGP) (see Figure 1.7). Since 1998, China has launched six national forestry programs: Natural Forest Protection Program (NFPP), Conversion of Cropland into Forest and Grassland Program (CCFGP), Sandification Control Program in the Vicinity of Beijing and Tianjin (SCPVBT), Shelterbelt Program in Three North area and along Yangtze River (SBP), Wildlife Conservation and Nature Reserve Development Program (WCNRDP) and Fast-growing and High-yielding Plantation Development Program (FHPDP). The first five programs function as forest ecological conservation and restoration programs to combat forest destruction and reduce its negative impact on the Chinese ecosystem. The last one – FHPDP – is

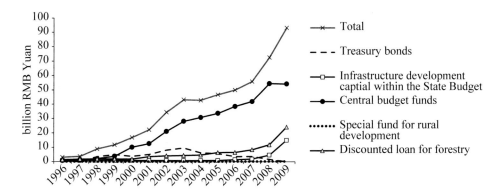

*Figure 1.6. Central investment in forestry from 1996 to 2009 (State Forestry Administration).*

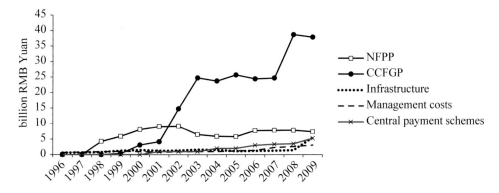

*Figure 1.7. Major directions of central investment from 1996 to 2009 (State Forestry Administration).*

a program for developing plantation for industrial use. It, also started at the same time, but and the program's main objective does not directly focus on protecting forest ecosystem but aims to ameliorate the shortage of timber caused by other forest protection programs. Planned investment for NFPP and CCFGP exceeded 316.2 billion Yuan (US$ 45.2 billion) in total by 2010 (Zhou, 2001). Different from traditional regulatory measures, the ecological projects provided subsidies to encourage local governments and forest owners to change their forest use practices and reduce their resistance in the implementation. In NFPP, local governments can receive compensation for their loss from harvest banning and investment for afforestation and forest management. In CCFG, local farmers are subsidized with grain and cash to convert their marginal cropland into forest and the local forestry bureaus can receive sapling subsidy from the central government.

Furthermore, a Forest Ecological Benefit Compensation Fund Program (FEBCFP) was established in 2001 with an annual budget of up to 3 billion Yuan (US$ 428.6 million), which has been raised to more than 5 billion Yuan since 2009 (State Forestry Administration, 2010a). The central government utilized the program to complement governmental regulations on forest use. Meanwhile, local governments also initiated payment schemes within their jurisdiction to target provisions of forest ecosystem services. Different from the usual forms of PES carried out in Latin America, USA and Europe, payment schemes in China have been largely dominated by governments and driven by the supply side – governments and communities who own forest resources (Li *et al.*, 2006c). The difference can be attributed to unique structural factors in the institutional setting of public benefit forest in China, including property rights, the distribution of funds, and responsibility sharing.

## 1.2 Problem description

With such a tremendous capital investment in these programs, questions have arisen about the environmental effectiveness and efficiency of these programs, especially compared with conventional "command and control" policies. Although the area and stocking volume of forest have increased rapidly and timber production has decreased significantly since the projects were launched (see Figure 1.3), the mechanisms behind this change have not been investigated thoroughly. The academic community and governments carried out a flurry of evaluations to measure the effectiveness and efficiency of the programs. However, most of the evaluations focused on the Natural Forest Protection Program (NFPP) and the Conversion of Cropland into Forest and Grassland Program (CCFGP), with little attention to the Forest Ecological Benefit Forest Compensation Fund Program (FEBCFP). Part of the reason for this imbalance in forest policy evaluation is that in the short term, NFPP and CCFGP provided a larger capital flow than FEBCFP and therefore plausibly had more significant environmental and economic consequences.

Results from various policy evaluations have shown that NFPP and CCFGP initially gave a strong punch to local forest industry and agricultural economy and produced positive environmental consequences. However, these evaluations also criticized the unsustainability of the environmental achievements, since the programs solely depended on short term national investment[3] without any formal payment institution for the period of post-NFPP or post-CCFGP (Zhang *et al.*, 2008). Compared to the NFPP and the CCFGP, the FEBCFP has a more stable financial source. It is supported by regular budgetary revenues, covering almost the whole country. And, in contrast to the national revenues of the other two, the FEBCFP succeeded in involving provincial and local governments to develop their own payment schemes for forest ecosystem services. In 2009, 25 provinces had established their own provincial payment schemes for forest ecological services (State Forestry Administration, 2010a). In addition many forest policy makers consider the FEBCFP an alternative to the NFPP and the CCFGP, and parts of forest under the NFPP have been transferred to the FEBCFP system. Although this looks promising, few empirical studies have evaluated the performance of these more regional and local payment schemes for forest services.

These ecological projects have been implemented in a top-down way. The heads of government at every level, instead of forestry officials, are asked to be responsible for achievement of project tasks. Specific departments in charge of projects are set up within forestry institutions. Forestry

---

[3] The duration of the NFPP is 10 years and that of the CCFGP was originally 8 years for forest regeneration and has been extended to 16 years. Although the central government required local government to provide additional funding proportional to central funding, few local governments completed the requirement in practice.

institutions at higher levels carry out frequent scrutiny on subordinate institutions. As a result, local officials often neglect the benefit of local farmers in order to accomplish the project task arranged by the central government. Some conflicts between local communities and project officials compromise the efficiency and effectiveness of the ecological projects.

Furthermore, some scholars also found that ecological projects looked more like a campaign in which the central government mobilizes a lot of resources and puts political pressure on local government to improve the environment in a relatively short period, but leaves decisive institutional arrangements untouched, such as the ambivalent role of forestry agencies. Just as Lang's predicted (Lang, 2002), we may find that after a few years the public salience of forestry protection in China declines, as attention shifts again to the political importance of economic development or the needs generated by some other crisis. It is possible that control over forests will again revert to the provinces, with predictable results in some of these provinces for a resumption of unsustainable exploitation. Although forest resources are still at stake and no apparent institutionalization is achieved in the implementation of ecological projects, this prediction might be a little too pessimistic as emerging payment schemes could provide new hope to institutionalize ecological projects in China's forest sector. As the former minister of SFA indicated (Zhou, 2002), the State on behalf of the entire society paid for ecological services from forests in the initial stage; subsequently it should change into a new role as coordinator to establish a rational compensation mechanism by legislation, taxation and redistribution of national income rather than to act as the direct builder, purchaser and compensator. But at present, gaps still exist between the basic compensation principle and implementation ideas in the Forest Law, and the actual rigid implementation without participation of stakeholders, which increases transaction costs and decreases effectiveness and efficiency of payment for forest ecosystem service.

## 1.3 Central research questions

This research aims to evaluate payment schemes for forest ecosystem services in China after they have been implemented for more than a decade. Rather than taking for granted the government's claim on the success of the payment schemes in protecting forests and providing ecological services, the research tries to scrutinize PES schemes with respect to ecological effectiveness, economic and livelihood impacts, participation of local people, and the interlinkages with forest tenure reform, by analyzing the performance of the payment schemes in different cases. Hence it tries to evaluate PES schemes in forests under the specific conditions of contemporary China

Effectiveness measures whether the schemes produce the ecological result that is intended. There are doubts from both international and domestic academia on the environmental effectiveness of ongoing forestry ecological conservation programs (Xu and Melick, 2007; Yeh, 2009). These conservation programs use a similar incentive mechanism as the payment schemes. Therefore, it is necessary to examine ecological effectiveness of the payment schemes, including both central and regional payment schemes for public benefit forest.

Furthermore, China's payment schemes usually not only target ecological conservation but also aim at improving local livelihood and reducing poverty. As a developing country with sizeable poverty, China still puts poverty reduction and livelihood improvement of local farmers on its political agenda. It is reasonable to assess the impact of the payment schemes on local livelihood, since a significant part of local poor people largely live on their forest resources and decoupling local livelihood from forest resources is decisive for the success of the schemes, especially its sustainability.

Successful payment schemes require participation, both to be effective as well as a value on its own. China's payment schemes, which are initiated by governments, are believed to ignore the engagement of local people in the implementation of the schemes (Xu *et al.*, 2006a). Therefore, assessing the participation of local people in the development and implementation of the payment schemes is important, also because this gives a specific Chinese "color" to these PES.

Finally, compared with the PES in other countries, where property rights on forest are relatively stable, China's collective forest regions are experiencing a tenure reform, which significantly reshapes the structure of property rights on collective forests, including public benefit forest. This makes it interesting to analyze the impacts (or potential impacts) of collective tenure reform on the effectiveness, livelihood, and participation of PES schemes.

Based on the objectives mentioned above, three research questions have been defined:
1. What have been the ecological and socio-economic effects of forest PES schemes in China?
2. To what extent and how have state and non-state actors (including farmers) participated in the design, implementation and evaluation of forest PES schemes in China?
3. How has forest tenure reform influenced the functioning and outcome of forest PES schemes in China?

## 1.4 Outline of the thesis

The next chapter offers a theoretical elaboration on China's ecological modernization and the payment schemes for forest ecosystem services and constructs an analytical framework for case studies. This is followed by Chapter 3 which provides an overview of the development of China's forest policies and more specifically forest resource management institutions, regulations and reforms that are closely related to payment schemes. In addition, the chapter ends with an introduction on research methods employed. This thesis uses case studies to analyze and evaluate payment schemes in different contexts.

The Chapters 4 to 6 are the empirical core of this thesis, as they analyze the payment schemes for forest ecosystem services in three different cases: Fujian, Guangxi and Liaoning provinces. In Chapter 4, focuses on the local payment schemes in Fujian Province, providing an assessment on their performance in terms of environmental effectiveness, livelihood impacts and participation. Chapter 5 deals with the case of Guangxi Zhuang Autonomous Region, which includes two types of providers for ecological services: the state-owned forest farms and villages. The evaluation demonstrates difference in performances of payment schemes under different forest ownership, management institutions and organization structures. Chapter 6 provides an analysis of forest policy and payment schemes in Liaoning province, with a specific emphasis on the impacts of collective forest tenure reform on the implementation of payment schemes. Chapter 7 provides a comparative analysis of the three case studies to identify similarities and differences among them and to analyze impacts of different institutional factors on the performance of payment schemes. The conclusions are generalized to other PES in China through a discussion on the representativeness of these case studies, and on the limitation of this research. Chapter 7 closes with recommendations for policy adjustment and future research.

# Chapter 2.
# Theoretical perspectives on payment schemes for forest ecosystem services

## 2.1 Introduction

In order to address the research questions, it is important to select concepts and theories suitable for conceptualizing payment schemes in ways that meet the research questions and objectives. This chapter first introduces Ecological Modernization Theory (EMT) and its development in order to provide a broad theoretical background. This is then followed by a discussion of the applicability of EMT in interpreting and understanding environmental improvements within the context of China. This section discusses the theoretical debates between various schools – debates which have informed this study. Next, there is a review of policy theories and approaches to evaluating environmental policy which is used as the basis for selecting the concepts used to build the conceptual framework for understanding the policy process behind payment schemes. The review briefly shows the differences and similarities between the various policy theories and evaluation approaches are used as a basis for selecting the theoretical stances and evaluative methods applied within this research. Finally, the evaluative framework is presented and each component is introduced in more detail.

## 2.2 Ecological Modernization Theory and China

Ecological Modernization Theory (EMT) is both a theory of social change and a political discourse. As a social theory, EMT analyzes how ecological rationality is increasingly changing the process of production and consumption in industrialized society. As a political discourse it is used to justify technical, economic, political and cultural transformation, following an ecological rationality – a path that is presented as being both a necessary and feasible way to deal with the crisis of modern industrial society. In the early stages, empirical evidence to support EMT came mainly from Western European countries, such as the Netherlands, Germany and the UK (Mol, 1995; 1999).

### 2.2.1 The development of Ecological Modernization Theory

In order to discuss the relevance of EMT to China, it is useful to present the history of this school of thought and its theoretical roots and to compare this to other social theories about environmental issues. In the beginning of the 1980s, scholars shifted their attention from explaining the roots of environmental crisis to developing an understanding of ongoing environmental reforms. The German author Martin Jänicke first employed the term ecological modernization to influence the German policy debate (Mol and Jänicke, 2009). The phrase was then taken up by the German sociologist Joseph Huber, who introduced it to academic circles, emphasizing the role of technological innovation in environmental reform. He proposed a seemly technocratic solution to environmental problems that was critical of the way in which the failing bureaucratic state handled environmental problems and saw a positive role for market actors, especially entrepreneurs and innovative companies, in environmental reform (Huber, 1991). From the late 1980s to the mid-1990s, the analytical focus of EMT shifted from technological innovation towards examining the roles of states and markets in environmental reform. Most of the EMT studies in this period focused on comparing the situations in industrialized countries. From the mid-1990s to the mid-2000s, EMT extended its theoretical scope into consumption studies and turned its sights to transitional countries in Central and Eastern Europe and emerging economies in East and Southeast Asia. Since the mid-2000s, EMT has extended its scope and started to theoretically relate to the emerging sociology of networks and flows (Mol and Spaargaren, 2005) which offered a new conceptual framework for empirical studies on global environmental governance and reforms. The geographical scope of its studies has also expanded and increasingly focuses on developing countries, especially African countries.

Debates with other theorists, such as deindustrialization or counter-productivity theorists, neo-Marxists, constructivists and post-modernists, have led EMT to establish its own theoretical foundation and evolve into a mature school of thought. EMT differs from the deindustrialization/counter-productivity ideas that inspired environmental movements in the early 1980s, which proposed to solve environmental problems by at least partially dismantling the systems of production. EMT argues that although design faults in industrial systems cause environmental damage, technological innovation and market mechanisms can also provide a solution to the environmental crisis provoked by industrial systems (Mol and Spaargaren, 2000). Early neo-Marxists, such as Allen Schnaiberg, argue that the capitalist organization of production in modern western society – the treadmill of production – is responsible for the environmental crisis. EM theorists dispute this and follow Giddens in treating industrialism and capitalism as separate institutional dimensions of modernity which leads them to focus more on the relationship between the industrial character of modernity and the environment (Spaargaren and Mol, 1992). These debates with counter-productivity theorists and neo-Marxists were formative for EMT's early stages, but they have since become outdated and largely irrelevant as a result of subsequent changes in theoretical focus and environmental discourses. The more contemporary debates have been discussions with relative constructivists and postmodernists about the material foundations of social theory, controversies with eco-centrists on radical and reformist environmental reforms, and challenges from recent neo-Marxists on social inequalities in environmental problems (Mol and Spaargaren, 2000).

EM theorists have also had to deal with postmodernists, who focus on deconstructing grand narratives. Postmodernist critics have looked at sustainable development and environmental problems which are increasingly defined in a globalizing context. They challenge the "real" and "objective" existence of environmental problems and see them more as "social constructs" framed by certain social actors in an arbitrary way. By contrast, EM theorists assert that environmental problems are "real", although they are also socially constructed through the specific framing processes of certain actors, according to their power and interests. Eco-centrists have been critical of EMT for advocating moderate proposals for environmental reform, rather than presenting a more radical vision of social change guided by ecologism. To a large extent, EM theorists agree that the cause of environmental crisis is deeply rooted in society's structure and culture and that the processes of production and consumption should be radically improved. However, they have a more optimistic attitude, based on the environmental improvements that have been made in industrialized countries in past decades and argue the case for realistic reform as opposed to radical societal changes. The ecological sphere and rationality is increasingly independent from other sphere and rationality. Fundamental alterations become not a necessary – at least not the only – solution to environmental problems. Although the primary focus of EMT is on environmental improvement, this does mean that it neglects other social issues, especially the link between social inequalities and environmental problems. EMT also draws on current neo-Marxist observations on the social inequalities that accompany contemporary processes of ecological restructuring. However, EM theorists take a different analytical perspective on the distributional effects of radical environmental reforms. EMT does not fully embrace the link made by neo-Marxists about the direct parallels between traditional class struggles and environmental struggles but is more inclined to the view that environmental struggles cross-cut economic interest lines and class divisions and should be regarded as a new category.

### 2.2.2 Ecological modernization as a social theory in China

More recent research in different types of countries (European, Asian and African; market-based economies and transitional economies; developed and developing countries) demonstrates the relevance (to different degrees) of EMT for understanding environmental reform outside the context of western Europe (Mol and Buttel, 2002; Mol and Sonnenfeld, 2000). Recent years have seen a wave of research focused on environmental issues in China, looking at environmental governance, industrialization, water pollution and deforestation. (Lang, 2002; Liu *et al.*, 2004b; Mol, 2006; Mol and Carter, 2006; Zhang, 2002; Zhang *et al.*, 2007). This body of research allows us to gauge the applicability of EMT in the context of China and the extent to which it can be used a tool for analyzing ecological compensation mechanisms in China.

In the late 1970s, China started its economic reform that transformed its patterns of production and consumption, and to a large extent this reform boosted the process of industrialization. Unfortunately China did not pay much attention to averting the environmental deterioration being experienced by industrialized western countries. Rapid economic growth was accompanied by growing problems with air and water quality and the loss of forests. At this stage, it was mostly natural scientists who were involved in carrying out environmental studies, including exploring the causes of environmental problems and trying to formulate solutions. Not surprisingly, these studies emphasized technological measures to combat environmental degradation. From the 1990s onwards, social scientists also became engaged in environmental studies, paying attention to the social factors behind environmental problems. They employed a range of different theoretical stances such as environmental justice, eco-socialism, social constructivism, deep ecology and ecological modernization in their analysis. This research introduced a sociological perspective into environmental studies, facilitated communication between the various theoretical views within the Chinese context, and developed China's environmental sociology – a discipline that is still under construction and far from maturity. Most of the theoretical foundations have not come face to face with each other through informed, critical debates (as has been the case in western academia), nor are there sufficient case studies to fully support the ability of competing theories to explain and provide solutions to China's environmental problems. Some scholars have even suggested the need for a distinctly Chinese environmental sociological theory to avoid inappropriately adopting social theories with roots in western society (Hong, 2010). This viewpoint questions whether ecological modernization is a suitable approach for understanding and analyzing China's environmental problems and whether the transition in China's environmental governance can be conceptualized as ecological modernization in the same way as its western counterparts. The following paragraphs review and discuss the development and application of EMT in China, to assess its suitability for this research. The contributions that EMT has made to environmental studies in China are also discussed, together with the reciprocal contributions that Chinese case studies can offer for the development of EMT.

The concept of ecological modernization was first introduced into China by several scholars in the field of environmental policy (He and Wu, 2001; Huang and Ye, 2001). Their work traced the development of EMT in western industrialized societies and suggested that it could offer a guide to environmental policymaking in China. Since then, the EMT perspective has been applied to a range of case studies on China's environmental reforms in different sectors. Zhang (2002) used EMT to analyze environmental management during the process of industrialization in small Chinese towns. Liu *et al.* (2004a) analyzed phosphorous cycles in China and demonstrated China's ecological restructuring in terms of material flows. Zhong (2007) developed an EM-based theoretical framework for studying institutional transformations in China's urban water sector and discussed reframing EMT to fit China's specific context. Zhang *et al.* (2008) took the Sloping Land Conversion Program in Ningxia, China as an example of ecological modernization, analyzing the impacts of participation and economic factors on the sustainability of China's ecological restoration program. Han (2009) evaluated renewable energy policy in China from the Ecological Modernization (EM) perspective. Mol (2009) and Zhang *et al.* (2010b) analyzed and assessed informational governance arrangements in environmental protection in China and one of the country's new informational governance instruments, the Environmental Information Disclosure Decree. Mol (2010) also undertook an innovative study of how the imperatives of sustainability shaped and patterned the 2008 Beijing Olympics, a global mega-event.

Alongside this empirical work some scholars have started to assess the appropriateness of EMT as a way of interpreting environmental reforms in China. This endeavor was encouraged by two major developments. First, China is making magnificent achievements in terms of economic growth, industrialization and in transforming itself from a centrally planned to a market-based economy. This is leading to the establishment of modern industrial institutions for production and consumption akin to those in western societies, even though China's level of industrialization is still ongoing and far behind that of western industrialized countries. This is a cornerstone of EMT's analytic assumptions and theoretical emphasis about the institutional dimension of modernity. Many of China's environmental problems are rooted in its emerging industrial systems and the solutions might, to some extent, be similar to those found in other modern industrial societies. In this sense, EMT could be a valid tool with which to analyze China's environmental reforms. Second, the acceleration of globalization is contributing to an increasing global interdependence in the political, economic and cultural spheres. One consequence of this is that the models, practices and dynamics of environmental reform are spreading from industrialized to industrializing countries through a range of influences: multilateral environmental agreements, transnational corporations, and environmental NGOs. As such it becomes not only theoretically intriguing but also practically necessary to examine the performance of these strategies, practices and measures of ecological modernization, which originated in western industrialized societies and are being transferred to new industrializing economies.

According to Arthur Mol's analysis (2006), an ecological restructuring of industrial systems is taking place in China in parallel with its economic reforms. In the early 1970s China established its structure for environmental protection – the National Environmental Protection Office (NEPO), predecessor of the Ministry of Environmental Protection (MEP) – and issued various environmental laws and regulations. Since then political modernization has taken place in China's environmental governance. Decentralization and flexibility offer new opportunities for local governments and environmental protection bureaus to develop environmental initiatives, strategies and institutional arrangements. Economic actors and market dynamics are playing a more active role in pushing for environmental reforms. Distorted prices for natural resources, such as water and forests, are being rectified and environmental costs are gradually being internalized into economic activities. Market demands are starting to motivate industrial sectors to take environmental interests into consideration and to restructure their production processes to make them more sustainable. Although China lacks Western-style environmental movements and NGOs, it does have special government-organized NGOs (GONGOs) which influence environmental policy through their expertise and closed networks. In addition, as a result of its openness policy, China is becoming more integrated with the international community, through trade, foreign assistance and bilateral and multilateral environmental negotiations, all of which contribute to and influence China's environmental policies and reforms. There is evidence that, in the process of China's modernization, ecological rationality is becoming increasingly independent from its economic counterpart and is beginning to influence the restructuring of production and consumption.

Although China's ecological modernization demonstrates some similarities with that of Western industrialized countries, there are also many differences because of the country's specific economic, political and cultural conditions. Environmental interests have only been partially institutionalized in production and consumption practices. Economic actors have not yet been subjected to sufficient and effective pressure to incorporate environmental interests into their production practices. In addition, environmental NGOs are relatively undeveloped in China and, to date, have only played a relatively marginal role in pushing for environmental improvements.

### *2.2.3 Ecological modernization as a political program in China*

Ecological modernization has also provided the basis for a government-led political program for greening industrialization. China's modernization project started at the beginning of the 20th century, but only stably unfolded after 1949 when the People's Republic of China (PRC) was founded. From the late 1940s to the late 1970s, the modernization and development of the economy was led by central planning and environmental concerns hardly influenced production and consumption processes. Starting in the late 1970s, market-based economic reforms which speeded up the modernization of the nation were implemented. Simultaneously, environmental issues acquired more importance with the setting up of an environmental protection office and the passing of laws and regulations for environmental protection. However, environmental interests were still marginal and the modernization process was overwhelmingly dominated by economic development. Equally, the "environmental state" was unsuccessful in controlling pollution and preventing ecological degradation. In 2005, the Chinese government made a major policy shift (in the 11th Five Year Plan – FYP – for National Economic and Social Development, 2006-2010). This marked a move away from a focus on "growth at any cost" towards a more balanced and sustainable growth pattern, under the "harmonious society" and "scientific outlook on development" policy frameworks. In 2007, the China Modernization Report 2007: a study on ecological modernization (China Center for Modernization Research, 2007) articulated the process of ecological modernization in China and presented a set of indicators to compare China with other countries. This report primarily used an economic-technological approach to ecological modernization and paid less attention to political modernization, civil society and public participation (Zhang *et al.*, 2007). The report's emphasis on science and technology (both in principle and in practice) can be seen as typical of the early stages of promoting environmental protection. This report received widespread media coverage that enabled it to exert leverage over political leaders and environmental policy makers. In October of the same year, at the 17th Party Congress, President Hu Jintao first made use of the term "ecological civilization" in his speech, listing it as one of the requirements for the late stage of China's modernization[4]. Gradually, this concept was taken up by the media, accepted by the public and more importantly, was employed by policy makers to initiate and reformulate environmental policies. In 2010, the guiding principles of China's 12th FYP (2011-2015) approved by the Communist Party of China's (CPC) Central Committee reiterated the concept of an ecological civilization and the importance of building a resource-saving and environmentally-friendly economic development model. Ecological rationality has been gradually built into China's modernization discourse and ecological interests have gained a central position on the national political agenda.

As mentioned above, the debates with other schools of environmental sociology have provided the momentum for the development of EMT. Similarly in China, EMT has faced criticisms and challenges from other theoretical viewpoints. These criticisms have helped to highlight missing points or imbalances that occur when attempting to transplant EMT from Western industrialized societies to China's context. China is still undergoing its modernization process, and economic growth and industrialization are still considered priorities for improving social welfare. Furthermore, the domestic environmental movement and NGO sector is relatively underdeveloped in comparison with the west, where they are often carriers of radical green ideology. Deindustrialization and counter-productivity have never been considered as a solution to China's environmental problems. By the same token, fundamental changes to the industrial system of production and consumption (proposed by neo-Marxists) are also not acceptable discourses. However, Chinese neo-Marxists do draw attention to issues of social and environmental justice and use these to critically analyze China's current environmental reforms. The political ecology orientation of neo-Marxism emphasizes the political dimensions of human-nature interactions, an issue that is relatively neglected by EMT. Huan (2007: 686) for example has criticized EMT as "a misleading way to understand and explain the complex dynamics for China's environmental

---

[4] This FYP strategically divides China's modernization into three steps: solving the problem of sustenance; realizing a comfortable life for society; and reaching the level of middle income countries and realizing modernization.

governance system", one which neglects active public participation and an environmentally aware citizenry. (S)He (Huan, 2007, 2010) also argues that the growth of the Chinese economy[5] has caused the current ecological crisis and that a fundamental change is needed to shift the current economic and social development model towards a model that prioritizes ecological growth. Huan also re-conceptualizes the concept of "ecological civilization" set out at the 17th Party Congress, developing a new radical vision which puts people's well-being ahead of profit-making and ecological sustainability in front of economic development, abandoning the economic growth model driven by large-scale investment and worldwide trade. Another criticism of the existing model made by neo-Marxists and political ecologists concerns the issue of environmental justice and the distributive effects of environmental reforms. While recognizing that substantial environmental achievements have been made in major cities and the eastern regions of China, Huan (2010) argues that the natural environment as a whole is bearing an ever increasing burden. Yeh (2009) argues that "ecological construction" projects (which is another name for ecological conservation programs and literally closer to the Chinese name) in western China are more than just environmental projects but need to be understood as political projects which create new rationalities of rule, new forms of subjectivity, and new economic and ecological practices. Thus ecological construction projects are being used to reproduce the nation and the state to create a new hierarchy of citizenship. These projects, she argues have the consequence of "marginalizing already politically and economically marginalized citizens, while producing only questionable environmental benefits" (Yeh, 2009: 892).

These criticisms raise two theoretical challenges for EMT. First, EMT asserts that it is not necessary to make fundamental changes to industrial production and consumption in order to create a sustainable society but that this objective can be realized by advancing science and technology, bringing market dynamics and economic actors into play, transforming the role of the nation-state and introducing better environmental governance. EMT would view China as still in the early stage of industrialization, (characterized by inefficient energy use, high emissions of pollutants and an overuse of natural resources) rather than as a growth-based economy that is addicted to profit-making. The problems are rooted in the country's imperfect market dynamics, weak institutional balances and limited public participation. The growth-led economic model may be exacerbating environmental problems, but it is not the root of them. Furthermore, exponents of EMT would argue that is wrong to label China's economy as a growth-led economy. While the prevalence of export-oriented industries and investment-driven economic growth could support this viewpoint, one can argue that this is just a temporary stage of Chinese industrialization and economic development and that domestic consumption will sooner or later become the major driver for economic growth. Of course, it must be admitted that there is still a risk that China will be trapped in this growth model, but even so, improving market dynamics, correcting price distortions for natural resources, improving the income distribution system, and adjusting the industrial structure would be a more feasible solution than fundamentally changing the industrial system and market mechanism.

---

[5] Huan (2010) defined a growth economy as a growth-oriented economy whose main objective is growth itself rather than the subsistence or wellbeing of human beings. Huan argues that there is an emerging Chinese "treadmill of production".

Secondly, one can argue that the criticisms of EMT for not paying enough attention to environmental justice and the distributive effects of environmental improvements are more related to China's specific context rather than to EMT itself. As Mol and Spaargaren (2000) indicate, environmental conflicts are distinct from traditional, social and economic problems and do not follow a predictable path of static opposing parties and interests. For example, farmers may be victims of soil erosion one day and victims of radical environmental reforms which limit traditional land use, the next. Clashes of environmental interests cross the traditional divides of economic groups and have to be treated distinctly. Although, in some individual cases, neo-Marxist studies can occasionally draw direct parallels between traditional class struggles and environmental struggles, they do not have enough theoretical or analytical value to describe or explain the environmental reforms currently underway in China. This said, there is a lack of adequate and effective participation mechanisms in China through which the interests of marginalized groups, such as farmers, can be represented in environmental policy making. As a result, the distributive effects of environmental reforms tend to follow existing political and economic divisions. It is not unusual for neo-Marxists to find many parallels between political-economic status and environmental interests. However, this does not mean that political ecology is the only legitimate perspective for analyzing environmental reforms in China. While it might be possible to use this approach to develop critical insights into power relations in China's rural areas, this carries the risk of neglecting the transition of environmental governance at the national and regional levels and its impact on local participation mechanisms, which in turn have had effects on environmental achievements. By contrast, EMT uses political modernization as a way of analyzing the distributive effect of environmental reforms. Rather than blaming environmental reforms for marginalizing certain social groups, EMT seeks to explore how environmental reforms function in the current institutional context and how political modernization changes environmental policymaking and reshapes environmental governance. This approach involves a systematic examination of participation mechanism in Chinese ecological construction projects, which should provide some insights into whether the EMT perspective can stand up to the theoretical arguments of the neo-Marxists.

Aside from these theoretical challenges, other schools (including neo-Marxists) have criticized the proponents of EMT for using unrepresentative case studies. EMT practitioners do not seek to claim that China has found a way of responding to all the environmental challenges it faces, but are interesting in documenting the first steps towards institutionalizing an ecological rationality in China's industrial complex and the emergence of a new form of environmental governance. As a result EM theorists have selected case study material that reflects these new trends in China's environmental reforms. Some of the early work along these lines was published in a special issue of Environmental Politics (2006) looking at "Environmental Governance in China". In this special edition, scholars (loosely) applied EMT to explain the institutional changes occurring in the economy, the political-administrative system and civil society, and assessed the actual transitions in environmental governance in four sectors (watersheds, energy, industrial transformation and genetic modification). This perspective, of an evolving system of environmental governance, shows increasing environmental improvements and institutional strengthening, especially in China's urban areas – a stark contrast to the apocalyptic environmental portraits. However there is still controversy about how to interpret environmental conflicts in China's rural areas where economic problems and the lack of local participation complicate the task of assessing environmental achievements and other social impacts. This provides a clear rationale for extending the application of EMT to examine environmental governance in rural areas. This thesis addresses this challenge by applying EMT to the forestry sector, which is often located in the poorest and most remote regions in China, where national environmental priorities interact with specific local conditions. The next section will discuss the appropriateness of EMT to understanding transitions in the governance of China's forests and whether it can provide a workable solution to problems in this sector.

## 2.3 Ecological Modernization Theory and forestry

"Compensation for ecosystem services" is a relatively new set of mechanisms designed to internalize the benefits of ecosystem services into the economic system. The emergence of such compensation mechanisms reflects two core features of EMT: economizing the ecology and political modernization (see Figure 2.1). These two core features are essential to understanding the logic of the development of China's compensation mechanism for Forest Ecosystem Services (FES) and for providing a frame of reference within which to evaluate different compensation policies.

### 2.3.1 Economizing ecology

The modernization process has resulted in economic rationality becoming the dominant rationale for organizing economic activity. Economic rationality implies that only goods that can provide economic benefit are monetarized and exchanged within the market. As a result, the pursuit of

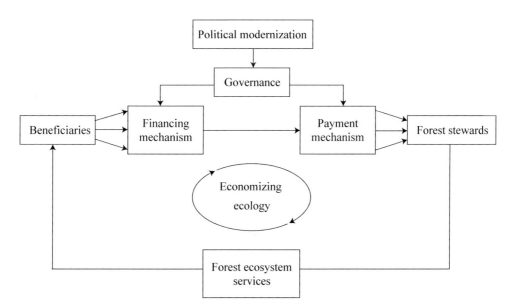

*Figure 2.1. The forest ecosystem services compensation mechanism seen from the ecologycal modernization theory perspective.*

economic rationality excludes the values of nature (e.g. the ecological benefits provided by a forest) which are not included or valued in the market system. EMT emphasizes the importance of internalizing ecological values within production and consumption processes, or economizing ecology (Mol, 1995). Economizing ecology means giving a market value or price to ecosystem functions and ecological benefits, bringing these functions elements that support and facilitate human life within the market system, so that they become valued and exchangeable. Through using the market mechanism, the external environmental costs of production and consumption are internalized.

Forests provide both productive materials (timber, fibers, land, etc.) and environmental services (water absorption, preventing soil erosion and silting and so on). These functions of forest are bound together and when the system is overused for a single purpose (e.g. timber extraction), this has a negative impact on the other functions. The process of industrialization has led to many forest resources being extracted and used as raw materials in industrial production. It has also led to great areas of forest land being cleared for industry and urbanization or for conversion into farmland. Thus natural forest resources have been transformed into industrial products for consumption and land (capital) for reproduction through the mediation of the market system which sends out price signals for the factors of production (raw material, capital, labor, etc.). During industrialization the market mechanism prioritizes the role of the forest as a source of productive raw materials and does not take the ecological functions and services provided by the forest ecosystem (as public goods) into account. This results in "insufficient provision" of forest ecosystem services, resulting in severe land degradation and other environmental problems. Over the past decade, a flurry of conservation innovations has emerged for forests (and other ecosystems) in the form of payment schemes and nascent markets for ecosystem services. In North America and Europe agro-environmental payments and private protection measurements now account for many millions of dollars of private and public expenditure (Scherr *et al.*, 2006), showing the extent to which ecology is being economized.

### 2.3.2 Political modernization: the role of states

While some economists believe that market instruments alone are capable of dealing with environmental problems, EM theorists consider that adequate governance structures are necessary, both as a direct instrument for making environmental improvements and as an institutional background to facilitate market instruments. But governance structures have often not proved sufficiently strong for dealing with environmental issues. The shortcomings of states in dealing with environmental issues highlight the need for two important transformations of environmental governance. First, state environmental policymaking has to change from curative and reactive to preventive; from exclusive to participatory; from centralized to decentralized and from over-regulated to inducing and stimulating forms of governance (Mol, 1995). Second, government has to shift part of its traditional responsibility to non-state actors such as individuals, communities, companies and NGOs. This allows non-state actors to obtain more opportunities to participate in environmental governance, giving rise to new forms of environmental governance, such as public-private partnerships, coalitions between businesses and environmental NGOs, private interest government, and other forms of "subpolitics" (Beck, 1992; Mol, 2006). However, these transformations do not follow a linear pattern of evolution "from an initiating government to a withdrawal-of-the-state", paralleled by a shift "from regulation to communication strategies". Rather, they are complex and pluralistic processes (Tatenhove *et al.*, 2000).

The development of public fiscal payments and market-based instruments for forest ecosystem services in China shows a number of these political modernization tendencies: a transition within environmental policymaking from reactive to preventive, from regulation to negotiation, from coercion by administrative power to stimulation by economic instruments. The emergence of local negotiations between the private sector and development projects sponsored by international organizations and NGOs displays the increasing engagement of non-state actors.

However, there are few studies that focus on or try to evaluate these payment schemes in China. The next section introduces major policy theories and how they have been applied to analyzing forestry policy, as the basis for developing an appropriate evaluative framework.

## 2.4 Policy theories and forestry policy analysis

A number of theories have been developed to analyze public policy, including behavioralism, rational choice, institutionalism, feminism, Marxism and normative theory (Marsh and Stoker, 2002). These theories are often based on different ontological and epistemological stances, take relatively distinctive positions in theoretical divides and demonstrate unique strengths and weakness for specific policy issues. Arts and Van de Graaf (2009) reviewed the use of different policy theories in the sub-discipline of forestry policy analysis and identified the top five theories used, or referred to, in the literature: policy networks, advocacy coalition framework, institutionalism, social constructivism and rational choice. This section first briefly summarizes these policy theories and then develops a theoretical framework based on the contribution that these make to the analysis of forestry policy.

Rational choice theory is a popular theory used in the field of political science. It assumes that individuals can make rational choices according to their preferences and that these choices form the foundations of political action. An advantage of rational choice theory is that it is able to build a micro-level model within the constraints posed by macro-level factors, which provides a dynamic link between the two levels and thereby serves as a plausible explanation for policy action. However, the theory does not say much about the formation of, and changes in, the preferences of individuals or organizations, which is an important aspect of human behavior. Rational choice theory was developed by those who believe that market mechanism is a better way to solve collective action problems than state intervention. They would point to Payments for Environmental Services (PES) as a typical example, which shows how rational choice triggers "the tragedy of the commons" (Hardin, 1968) and policy initiatives should be developed to deal with both market failure and state failure.

Institutional theory or institutionalism, places less emphasis on the agency of actors than rational choice theory. It sees the public policy process as based on interactions between structural power – "institutions" and actors. It argues that an examination of policy processes should focus on the organizational contexts, which are replete with established norms, values, relationships, power structures and standard procedures (Hill, 2005). This theoretical stand often underpins research into PES, and by focusing on institutional factors is often employed to design PES programs that are effective and efficient (Scherr *et al.*, 2006). However, institutional theory experiences difficulties in explaining policy change. Therefore attempts have been made to combine institutionalism with rational choice theory in order to provide a more complete explanation of policy process that provides a better balance between agency and structure. In addition, an "institution" is a very broad concept, which ranges from formal constitutions and rules to informal conventions and culture (and so can also include ideas and discourses). This conceptual flexibility creates opportunities for researchers to understand various institutional components of existing PES from numerous international experiences.

Combining perspectives from these two theories, the institutional analysis and development (IAD) framework focuses on identifying an "action arena" and analyzing the factors that influence it, including the rules that individuals use to order their relationships, the attributes of the physical world and the nature of the community within which the arena is located (Ostrom *et al.*, 1994). It uses this as a point of departure from which to explore the resulting patterns of interactions and to evaluate policy outcomes. This framework has been applied in common-pool resources studies, especially for community forest management. The IAD framework has developed theories and models that examine how forest use practices are monitored and how local institutional arrangements influence the behavior of forest users, with different ecological consequences. These theories and models provide valuable institutional building blocks for establishing PES schemes (Landell-Mills and Porras, 2002).

Social constructivism takes a distinctive ontological and epistemological stance towards public policy analysis. It argues that social problems are not neutral or objective phenomena but are interpretations of conditions that have been subjectively defined as problematic (Ingram *et al.*, 2007). The uncertain nature of environmental issues gives a great deal of freedom to socially interpret or construct a "problem" and its solution. This is especially true for forest ecosystem services. Researchers have criticized policymakers for initiating projects or banning logging to improve water conservation on the basis of their beliefs, rather than on scientific evidence (Kaimowitz, 2004). Some research results have also demonstrated the contested nature of interpretations of scientific evidence: to function well a PES should clearly define what is being bought; the less rigorous the scientific basis of a PES scheme, the more vulnerable it is to the risk of buyers questioning its rationale and abandoning payments (Wunder, 2005). Several recent pieces of research have applied discourse analysis as a way of critically examining changes in forest policy and the emergence of PES schemes (Arts and Buizer, 2009; McAfee and Shapiro, 2010; Pereira and Novotny, 2010; Van Gossum *et al.*, 2011). Discourse analysis is a paradigm that uses ideational and symbolic systems and orders to understand and describe the social world. In discourse analysis, the knowledge-driven and meaning-searching "homo interpreter" replaces both the rational "homo economicus" and the norm-driven "homo sociologicus" as the starting point for explaining social practices and societal change (Reckwitz, 2002). Discourses are seen as both the outcome and the medium of human action. People and groups form discourse coalitions which give discourses traction in political processes (Hajer, 1995). Discourse analysis aims to understand both how human actors construct discourses and how "existing" discourses mediate this process. Discourse analysis has also drawn on insights from social constructivism to further develop the theories of institutionalism. Arts and Buizer (2009) have introduced discursive-institutional analysis, using it to explain emerging discourses about sustainability, biodiversity and governance within the global forestry regime. Their analysis shows how these new ideas and meanings have been institutionalized over time.

Policy network theory is an approach devised to better explain and describe the shift in the policy domain from "government" to "governance". This approach emphasizes informal, decentralized, and horizontal relations, rather than formal, centralized, and hierarchical government agents (Adam and Kriesi, 2007). It analyzes the policy process through specific network configurations, in which actors or stakeholders cluster together around specific shared interests and/or beliefs relating to a policy.

The Advocacy Coalition Framework (ACF) has much in common with the policy networks approach and sees the policy process as involving an advocacy coalition (Hill, 2005; Sabatier, 2007). These two theories are currently very popular in forest policy literature, ranking first and second respectively (Arts and Van de Graaf, 2009; Lazdinis *et al.*, 2004; Singer, 2008). They offer a useful way to analyze the clustering of interests in forest policy, especially when the forestry policy domain is increasingly being restructured by processes of globalization and decentralization. In the same way, PES schemes are also increasingly embedded in an international context, related to global environmental issues, where the networks approach can have a theoretical and explanatory advantage (Swallow *et al.*, 2007).

This thesis aims to develop an analytical framework to better understand and describe PES in China, seek ways of evaluating their effectiveness and explore the relationship between different evaluative elements. To this end, this framework does not directly follow any of the theories discussed above, but takes insights from different theories to build its different components. Overall the framework draws more heavily on institutionalism, as the institutional context delimits the arena and possibilities for the various actors involved in the payment schemes. Institutional factors play an important and perhaps even dominant role in influencing the local performance of PES. However, this does not mean that they are always a decisive influence on performance. Institutional rational choice theory and the IAD framework both offer insights into how interactions between actors play an important role in determining policy performance. The attributes of the physical world, the nature of the community and the formal and informal rules that create incentives and constraints are all other important factors that need to be considered when examining the performance of payment schemes. For analytical convenience, we re-integrated these factors and embedded them in the framework of this thesis. The framework first identifies an institutional setting that includes external rules. These external rules interact with each other and form a basic setting for emerging payment schemes. Yet because of the time-space dimension, this setting is also constantly evolving (within certain parameters). The payment schemes are embedded in this dynamic setting: they are parts of an institutional context in which actors interact and forestry practices are carried out. At the same time, they impact upon external institutional factors.

The social constructivist approach has only relatively recently been applied to understanding and describing the dynamics of forestry policy, and is not incorporated within this analytical framework. This thesis concentrates on policy evaluation, with the focus being on finding a feasible way to evaluate the material and interest oriented dimension of forestry practices, rather than attempting to explore the roots of institutional change in the forestry domain. However, discourse analysis can be used to describe the institutional setting for payment schemes and can offer useful insights into the dynamics and mechanisms involved (beyond interests and power).

These insights will be only limited within the description of the institutional setting without entering into the framework.

The policy networks approach provides useful insights into the relationships between farmers, village committees, and forestry bureaus. It is also useful for considering the current debate over the extent to which a transition from government to governance is emerging in the Chinese forestry sector. More specifically can the payment scheme mechanisms be understood as part of a process of transformation from formal, centralized, and hierarchical government agents to informal, decentralized, and horizontal governance structures? The answer is rather ambiguous. The general trend in Chinese forest governance is towards decentralization with the state becoming divorced from the previously state-owned forest enterprises and farms, forest management is shifting from collectives to individual farmer households and timber prices are being increasingly determined by the market. However, the "public benefit" forest is largely influenced by large-scale governmental ecological restoration projects which show a centralized management style. The government is still the largest sponsor, implementer, and inspector of these schemes. China's forestry sector is at a historical turning point and no clear answer can be recognized. Nevertheless, the policy networks approach can still provide insights into the interactions between policy communities within the payment policy domain, especially to offer a normative reference frame for the participation mechanism of existing payment schemes. Although the payment schemes in China are dominated by government, the foundation of forest governance in China – tenure structure – is changing. Thus there is an inherent contradiction between centralized ecological management and decentralized forest tenure structure, which is likely to become more pronounced in the future. The policy network approach can provide a unique lens for examining the role that different policy communities play in the policy process of payment schemes and whether participation goes some way to resolving this contradiction.

Policy theories can provide assumptions, mechanisms, and expectations related to the policy process. These theoretical parts can provide a theoretical foundation, perspectives, insights, and a lens for policy evaluation. However, an analytical framework for policy evaluation should be also supported by a set of methodologies to measure and evaluate public policy. Policy evaluation approaches differ from policy theories in that they focus on developing a framework or paradigm to examine and evaluate policies. This kind of examination and evaluation usually does not seek to explore the mechanisms or dynamics underlying the policies, although some evaluation approaches do integrate such inquiries within their framework. However the primary mandate of these approaches is to examine policies from different perspectives, even if the policy process remains a "black box". The following section provides a brief summary on approaches to evaluating environmental policy (summarized in Table 2.1) and selects suitable perspectives for measuring and evaluating the performance of payment schemes.

## 2.5 Approaches to evaluating environmental policy

There are different classifications for approaches to evaluating environmental policy, according to different epistemological stances, disciplinary orientations, research paradigms, etc. Each approach serves as a unique window through which to examine environmental policy processes. It is not my intention here to develop a thorough typology of these approaches. The distinction

Table 2.1. The main approaches to evaluating environmental policy and the differences between them.

| Approaches to evaluating environmental policy | Phase in policy cycle | Main objectives and features |
|---|---|---|
| Formative evaluation | Preparation and implementation | Aiming to improve policy implementation |
| Environmental impact assessment | Between planning and implementation | Providing alternatives for reducing negative impacts |
| Goal-free evaluation | After implementation | Detecting side-effects of a policy |
| Experimental and quasi-experimental methods | After implementation | Establishing causal relationship between policy and its effect |
| Economic appraisal and evaluation | In planning and after implementation | Assessing economic efficiency or feasibility of policy |
| Theory-based evaluation | In planning and after implementation | Carrying out policy evaluation based on a theory |

between the different approaches lies in the perspectives and methods that the approaches use to break down or select the components of public policy in order to construct a frame of reference for evaluation. This section first briefly reviews the most widely used approaches and theoretical stances on evaluating for environmental policy and then goes on to set out the evaluation approach to be used in this thesis.

Formative evaluation is an action-oriented evaluation method. It involves examining the implementation of policy and assessing the policy process and its organization, with the aim of improving and optimizing policy programs. It can be carried out when policies are being prepared or implemented. It focuses on the development of policy and its functioning mechanisms, especially when the policy is subject to frequent change (Patton, 1994).

Environmental impact assessment is an institutionalized policy instrument for the *ex ante* evaluation of the environmental impact of policy programs or projects that is designed to identify possible alternatives for reducing negative impacts. Generally, the evaluation involves several steps including a scoping process, identification of the impacts, assessment of value, estimation of likelihood and synoptic predictions of the impacts.

Traditional evaluation is generally goal-orientated and seeks to examine the consequences of a policy against its objectives. However, this method often neglects any side-effects of the policy, which can often occur when implementing environmental policy. Goal-free evaluation offers a possible solution to the shortcomings of traditional evaluation (Scriven, 1991). In such a study an independent evaluator examines the broad range of policy effects, rather than just focusing on the goals of the program. A series of evaluation criteria are formulated, drawing more on the perspective and experiences of the target group and other stakeholders than on those of the policy makers. The evaluator then uses these criteria to assess the effects of the policy. This approach shares the same philosophy as needs analysis, which focuses on comparing policy and its expected outcomes with the needs of the target group. Needs analysis is only used as an *ex ante* method during the formation stage of policy, compared to goal-free evaluation which can only be applied *ex post*.

Campbell and Stanley (1963) describe experiments as "the portion of research in which variables are manipulated and their effects upon other variables observed". An experimental approach can be used as a means of establishing the causal relationship between a policy and its effects. This method differs from other evaluation methods, in that it requires a strict experimental design to guarantee the validity of evaluation and avoid falsely claiming a causal correlation between a policy and its outcome. This method is largely affected by laboratory research of nature science. The experimental method can provide valid and reliable evidence about the effectiveness of a policy (compared with other policies), or through randomized control trials that reduce the influence of extraneous factors. However, randomization cannot be always achieved in reality, since it is often difficult to practically implement a policy experiment and the size of sample data is too small to be valid. The quasi-experimental method loosens the limitation on randomization that allows comparisons to be made between experimental and control groups. Here, various methods are employed, such as time series (which can also controlled for before and after designs), equivalent sample design, counterbalanced designs, etc. (Crabbé and Leroy, 2008). Although experimental and quasi-experimental methods have advantages when strictly testing the causal relationship between policy and its effects, especially environmental effects, which are influenced by many extraneous factors, randomization runs the risk of eliminating the explanation of the context on the effects of the policy. In some cases, the context has more impact than the policy intervention. In addition to this, experimental and quasi-experimental methods do not allow any opportunity to study the side effects of a policy.

Economic appraisal and evaluation is often used for analyzing and assessing the economic efficiency or feasibility of policies. Cost-effectiveness analysis (CEA) and cost benefit analysis (CBA) are the most commonly applied economic evaluation methods. CBA seeks to examine all the costs and benefits brought about by a policy intervention, which are expressed as a monetary value, with the costs and benefits being compared to determine the desirability of the policy. When evaluating environmental policy many major costs or benefits (e.g. soil erosion or cleaner air, etc.) do not have a market price and are not exchanged on the market. As a result, it is problematic for CBA to integrate these costs and benefits into analysis. A number of methods (such as travel cost methods, replacement cost methods, and contingent valuation, etc.) can be used to estimate the monetary value of these costs and benefits. However, there is still much controversy over the use of these methods in assessing non-monetized goods and services. Unlike CBA, CEA does not attempt to measure the effects of policy interventions in monetary terms, but seeks to directly compare the achievements of different policy options at the same cost in order to identify the most efficient option. Although CBA and CEA provide a clear evaluation of the efficiency of policy, they do not take its distributive effects into account, how the costs and benefits are divided among different groups or individuals. This issue highlights that aggregate efficiency alone provides no guarantee of the feasibility or acceptability of a policy.

Theory-based evaluation[6] consists of an explicit theory or model of how a policy causes the intended or observed outcomes and an evaluation that is, at least partly, guided by the model (Rogers *et al.*, 2000). Realistic evaluation is one type of theory-based evaluation which assumes that effects are the product of the interaction between a mechanism and its context (Pawson and Tilley, 2003). It defines a series of concepts to uncover causality, such as context, mechanism and generative causality. The evaluation approach examines whether the mechanism can generate desired outcomes under a certain context. Realistic evaluation relies on a natural scientific approach to causality rather than on the traditional approach which relies more on serial observations or the perception of the researcher. Methodologically, realistic evaluation is articulated through interviews with a purposive sample of (a limited number of) stakeholders. This is in contrast to other evaluation methods for public policy which tend to employ a process of building consensus among a wide group of stakeholders (Blamey and Mackenzie, 2007). In this sense it is more suitable than experimental methods for unraveling the mechanisms underlying policy (Crabbé and Leroy, 2008).

It is not possible to combine all the dimensions offered by these different evaluative approaches into one analytical framework. The dimensions selected for evaluation should fit with the research topic and the purpose of the research. In China, payment schemes for forest ecological services have been in place for about ten years. Although they are environmental programs which objectively focus on protecting and conserving the forest ecosystem, their implementation has side effects, particularly negative economic and social impacts on local people. Traditional goals-based evaluation does not offer a suitable framework to catch these side effects but the goals-free evaluation method does have the potential to identify the economic and social impacts of these schemes. At the same time, the policy on payment schemes is constantly changing, with new payment schemes emerging. In this regard, the evaluation of payment schemes is formative, not only examining the implementation of the payment schemes but also providing suggestions for improving the schemes. When evaluating policy, it is important to establish the causality (if any) between a policy and its effects. Although this is usually addressed through methodological issues such as research design and sampling strategies, an appropriate conceptual framework can also contribute to identifying any strong causality between a program of payments and its effects. To achieve this, the evaluation will be based on a social theory about the implementation of payment schemes. This is a theory-based evaluation which emphasizes the interaction between a payment mechanism and the context in which it operates. The institutional setting, local physical conditions and the characteristics of the community are all important aspects of the context in which payment mechanisms operate. A payment mechanism can cause changes in local practices of forest use. Therefore the evaluation needs to not only include a measurement of the performance of the payment schemes in terms of local forest practices, but also to explore the relationship between the institutional setting, the payment schemes and forest use practices. The analysis of this relationship can help identify a reliable causality between payment policies and established effects. An additional purpose of this research is to respond to theoretical criticism of EMT from other schools, which

---

[6] There are different terms used to refer to essentially similar evaluation approaches. These include theory-based, theory-driven, theory-oriented, theory-anchored, theory-of-change, intervention theory, outcomes hierarchies, program theory, and program logic (Rogers and Weiss, 2007).

argue that EMT's perspective neglects the economic consequences of environmental problems and solutions. Therefore, the evaluation of the economic impact of payment schemes also needs to use cost benefit analysis to identify the economic impacts on local people.

## 2.6 An evaluative framework for payment schemes

Implementing PES can have different implications. On the one hand, PES is intended as an instrument to promote conservation and sustainable management of natural resources. It is an innovative tool that is widely employed in developed and developing countries to strengthen the supply of environmental services by channeling financial resources from beneficiaries to producers. On the other hand, PES also contains a neoliberal environmental discourse, which takes the premise that "the natural environment can best be safeguarded by valuing and managing "nature's services" as tradable commodities" (McAfee and Shapiro, 2010: 2).

PES differs from traditional policy instruments in several respects. First, PES emphasizes economic incentives as a way of achieving environmental improvements, in contrast to traditional regulatory policies which rely more on command and control. Experiences around the world demonstrate a failure of these command and control methods to actually achieve environmental improvements. PES, which is being introduced and applied in the field of forestry conservation and restoration, makes it possible to complement or replace ineffective traditional policy instruments by offering incentives for forest protection. PES offers economic advantages: it provides economic compensation to providers of ecological services and ameliorates or eliminates conflicts between economic and ecological interests. Second, PES has a different governance structure than traditional policy instruments. It encourages consultative, negotiated solutions to environmental problems rather than an arbitrary, coercive method, which usually has negative social effects. This requires a governance structure that distributes the responsibilities and benefits and which can implement and monitor forest use and management in a decentralized way (as opposed to the centralized approach typical of traditional administrative institutions).

PES can be categorized according to the source and method of payment. In terms of the sources there are two categories: public fiscal payment schemes and market-based payment schemes. Public fiscal payment schemes involve the government taking responsibility for closing the gap between private and social costs in environmental issues and increasing the supply of environmental services by paying for them as a public good. Proponents of market-based payment schemes claim that public fiscal payment schemes are inefficient in their implementation and have high monitoring costs. They argue that the market mechanism should be introduced to reduce transaction costs for environmental services and increase their efficiency. However, there are a lot of preconditions for running efficient market-based payment schemes. These include the existence of a trustworthy forest tenure system, identifiable and marketable environmental services and sufficient social capital to support the market mechanism. In the forest sector, PES are made for a range of ecosystem services including: watershed services, biodiversity and carbon sequestration at different scales, from global to local. Table 2.2 demonstrates the main payment schemes for forest ecological services and their characteristics.

Despite being widely applied, the performance of PES has not yet been carefully examined in China. In China some information about the growing number of PES schemes being introduced

Table 2.2. Main payment schemes for forest ecological services around the world.[1]

| Category | Approaches | Stakeholders | | | Instruments | Pricing |
|---|---|---|---|---|---|---|
| | | Buyers | Mediators | Suppliers | | |
| Public fiscal payment scheme | Governmental purchase | Government | Governmental agency | Private land owners | Purchase contracts | Individual negotiation |
| | Governmental subsidy | Government | Governmental agency | Private owners and communities | Grant contracts, conservation easements | Single fixed standard, negotiation or auction |
| Market-based payment scheme | Conservation contract trading | Land trust | – | Private owners | Conservation easement | Direct negotiation or donation |
| | Equity finance | International organizations, stock market | International environmental fund | Private entrepreneurs | Risk investment, stock investment | Direct negotiation |
| | Eco-service credit exchange | Government, NGOs, enterprises, consumers | Exchanges, brokers, and other market actors | Eco-service project developers | Eco-service credits and their financial derivatives | Exchange pricing, over-the-counter, etc. |
| | Eco-labeling | Consumers | Eco-labeling certification institutions | Companies and producers | Commodities with eco-labeling | Competitive pricing in market |
| | Eco-service fee | Direct beneficiaries (enterprises or individuals) | Government, NGOs, local communities | Forest owners | Eco-service fee | Governmental pricing, direct negotiation |

[1] Summarized from cases of international payment schemes.

in the country has been published recently, but there is little detailed information about the implementation of the PES schemes and their performance (Chen, 2006; Sun and Chen, 2002; Xu et al., 2006a; Zhang et al., 2008; Zheng and Zhang, 2006). Furthermore, the transformation of PES from an idea into an institutional practice in China is still an ongoing process: the new payment schemes emerged to change forest management practice; PES had to be shaped and framed as a viable payment scheme; the rules needed to be formalized and implemented in the local context; traditional authorities lost their absolute control over natural resources, attributing to the empowerment implied by PES and forest ownership change. It is necessary to develop an evaluative framework to understand the policy process behind China's payment schemes and evaluate them in the context in which they are located. At the same time PES provides an excellent avenue for examining the applicability of EMT as a way of interpreting China's environmental reforms in rural areas. The following paragraphs provide an overview of an evaluative framework which could fulfill these aims. This is followed by a detailed description of the individual components of the evaluative framework and a justification for selecting them.

Based on the above summary of approaches to evaluating environmental policy, this study developed a goal-free, theory-based, formative evaluation framework with economic evaluation components. The evaluation aims to examine the performance of payment schemes and to formulate suggestions for improving these schemes. It is closely related to the research questions of whether payment schemes are effective and why they succeed or fail. The payment schemes have been in place for about ten years during which time forest tenure reform (and other forestry management reforms) has been evolving, changing the context in which the schemes operate. While some performances looked apparent to evaluate, some challenges and problems have emerged in implementing the payment schemes. A formative evaluation can be used to measure the effectiveness of the payment schemes and to explore possible solutions for future implementation. A decade after being implemented, the payment schemes have had various unforeseen consequences for the local environment, economy and society. Thus the evaluation cannot be restricted to just the original objectives of the payment schemes but should also explore the policy's side-effects. One of the most notable side-effects of the schemes has been the negative economic effects on local forest owners. A Cost-Benefit Analysis (CBA) component has been employed to try to better understand the economic impacts on local forest owners. In addition, in order to explore the mechanisms linking the payment schemes with these outcomes, the evaluation employs a theory-based method, explained below.

Overall, the evaluation framework focuses on three aspects of payment schemes: policies, practices and performance (Figure 2.2). The analysis of policies focuses on the relationship between the institutional setting and payment schemes. The implementation of payment schemes (via the institutional setting) affects forest use practices, which is the core element of the evaluation. Forest use practices are affected by payment schemes and this in turn influences the formulation and implementation of subsequent payment schemes. This research aims to evaluate the dynamics of this process from policies to practices. Three dimensions were selected to explore and evaluate the process: environmental effectiveness, economic impacts and participation.

### 2.6.1 Institutional setting as the context

When exploring policy intervention it is important to understand the institutional setting, which has a profound influence on the likely effectiveness of any policy intervention. Any comprehensive analysis of payment schemes should include the institutional setting within its framework. The institutional setting can be defined as the set of formal and informal rules within a specific policy domain, which enables or constrains the formation, implementation and transformation of a certain policy instrument. The institutional setting for payment policies consists of various elements such as the location of the forestry authority, formal regulations on forestry policies, organizations, the structure of tenure, development strategy, the local political culture and administrative style

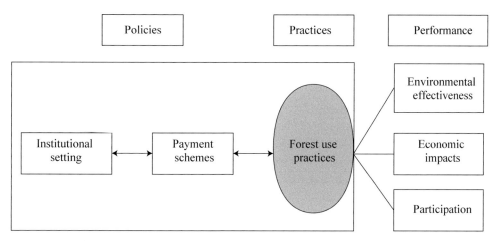

*Figure 2.2. Evaluation framework for payment for environmental services: policies, practices and performance.*

etc. Although the institutional setting can be regarded as a relatively stable environment, some dynamic factors (in this case reform of forest tenure) can transform the institutional setting during the implementation of a payment scheme. As such the evaluation of the payment schemes should consider both regionally-specific factors and the dynamic factors of institutional setting.

### 2.6.2 The mechanism of payment schemes

The introduction of payment schemes into forestry protection policies can be regarded as a process through which the Chinese government has built some of the core principles of sustainable development into its forest policies, such as equity, participation, the precautionary principle and policy integration. It can also be interpreted as a practice of ecological modernization, the introduction of incentive-based measures, instead of solely relying on coercive policy instruments.

A PES mechanism usually involves three sets of actors: the beneficiaries and suppliers of the ecosystem services and the mediators who try to distribute the money from the beneficiaries to the suppliers who are delivering the services (by following management plans, etc.). In China, payment schemes for FES are largely managed by the government, which plays the role of mediator in channeling the financial flow from beneficiaries to suppliers and also monitors the provision of these services (Figure 2.3). The government also represents the beneficiaries of the payment schemes. Beneficiaries are rarely singled out as independent participants in the formation or implementation of payment schemes. However, governmental agencies take different positions in the policy process and this shapes different payment schemes under various institutional settings. Third, although payment schemes signify an innovative instrument in forest governance, they are not the only mechanism in place. They coexist alongside traditional coercive measures, thereby creating a hybrid governance structure in the forestry sector. EMT is employed to understand and describe the characteristics of the payment schemes and the underlying dynamics within the mechanism.

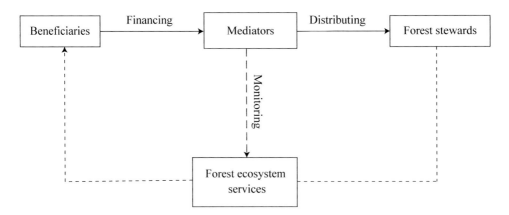

*Figure 2.3. The forest ecosystem services mechanism: a simplified view.*

### 2.6.3 Forest use practice

Forest use practice refers to the ways of using and managing forest resources (Figure 2.4). One of the key aims of payments for FES is to alter these practices on the ground, so that they correspond more with sustainability goals. Payment schemes provide constraints and incentives for forest use practices, which occur within specific local contexts that consist of the natural physical conditions, the local economic and social situations, and the characteristics of the local communities. Payment schemes are intended to alter practices and encourage local people to routinely adopt new practices in their daily lives. The policy interaction pattern or narrow participation mechanism is stabilized between different actors. These changes in forest use practices will impact upon the environment, economy and society. As these changes take shape they may even lead to a reshaping of the payment schemes (in the long run). In this sense forest use practice determines whether payment schemes achieve their desired outcomes – they are the causal link between policy and performance. Such practices not only impact directly on forest resources but also influence the mobilization, deployment and management of resources by different actors. For example, timber harvesting is influenced by many policies at the macro level including logging quotas, payment schemes, and forestry industry development plans. At the local level logging will be influenced by a range of factors, such as monitoring efforts, willingness to comply and the likelihood of punishment or sanctions. The attitudes of local managers will also influence the logging patterns adopted by local actors.

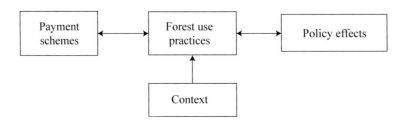

*Figure 2.4. Forest use practices.*

## 2.6.4 Evaluating the performance of FES payment schemes

When seeking to evaluate the performance of FES payment schemes a range of questions emerges. How should one measure the performance of such payment schemes? Should such an evaluation cover environmental consequences, economic and social impacts, or other things as well? What should be examined and evaluated – the actions of government, of local management, farmers' behavior, or the whole policy process? What is the nature of performance evaluation in this research? Should the evaluation look at the consequences of a payment scheme that lie beyond the original objectives? How does one integrate economic evaluation into the equation?

The evaluation of the performance of a policy can indicate how well or badly the policy has been carried out. This evaluation can be based on different dimensions, such as effectiveness, economic efficiency, distributional equity, social impacts, accountability, fiscal sustainability, etc. Environmental policies differ from most other public policies in that their ultimate target is not to influence patterns of human behavior, but to enhance the state of the biophysical environment by achieving changes in human behavior (EEA, 2001). The evaluation of payment policies will therefore focus both on the impact they have in changing the behavior of forest users and other stakeholders as well as the quality of forest ecosystem services.

The European Environmental Agency (EEA) (2001) has developed a framework for environmental policy evaluation which identifies a number of key elements that influence or mediate the influence that a policy measure has on human behavior and ultimately on the environment. By comparing these different elements it seeks to answer a broad set of evaluative questions. The framework they set out is an *ex post* summative, goal-free, method-based policy evaluation. This framework provides a starting point to formulate questions that can be used to evaluate payment schemes. These might include: Are the objectives of the payment policies justified in relation to needs of multiple stakeholders? Are the payment policies effective in providing the desired FES and against the cost? Do the outcomes and outputs meet the objectives of different stakeholders? Environmental effectiveness and economic efficiency are two central aspects of the evaluation. However, this framework only offers a static comparison between the objectives of a policy and its outcomes. What is the relationship between the institutional setting and the payment schemes? How do the payment schemes change forest use practices and thereby achieve the desired environmental effects? What is the pattern of interactions between the different actors within the local context under the payment schemes? The pattern of interactions can influence changes in forest use practices. Examining the patterns of interaction can not only unveil the functioning mechanism of payment schemes, but can also help to identify how to make them more effective and efficient. Therefore, the evaluative framework in this study is built on the policy process theory of payment schemes (understanding the interactions between the institutional setting, payment schemes and forest use practices), rather than on a method-based evaluation such as that proposed by the EEA, which only focuses on consequences of environmental policy.

EMT offers useful insights for understanding the policy process of payment schemes and constructing the evaluative framework. The theoretical foundation for developing payment schemes is to "economize ecology". The extent to which FES are economized signifies how effectively the providers of services can be compensated. Furthermore, the process of economizing indirectly relies on the monitoring and coordination mechanism of the payment schemes. Therefore, the efficiency and effectiveness of payment policies is closely related to the process and degree of economizing forest ecosystem services. The concept of political modernization not only refers to the changing role of the state, but also to an increase in participation in policymaking and implementation. Payment policies for FES differ from traditional governmental regulations in that they create opportunities to bring both forest stakeholders and beneficiaries into negotiations about the provision of FES. Such participation also has an impact on the accountability and legitimacy of payment schemes for FES and ultimately influences their efficiency and effectiveness. As a consequence, participation is an important aspect that should be built into policy evaluation. The following section elaborates further on the three main evaluative aspects.

1. Environmental effectiveness. Environmental effectiveness can be judged in terms of the extent to which the expected objectives of a policy measure are achieved (EEA, 2001). The main objectives of environmental policies are usually to improve environmental conditions or to reduce ecological risks. The effectiveness of a measure can be judged by comparing its environmental effects with its intended objectives. To measure the effectiveness of the payment schemes, it is necessary to explore the causal relationship between a scheme and its effects. The payment schemes are a policy intervention intended to influence forest use practices and thereby create environmental benefits. It is important to measure these since the beneficiaries are more likely to be willing to pay for these services if they can clearly see the level of ecological services that the payment schemes generate. However, most of the PES programs in China do not include a clear and explicit framework for monitoring or evaluating their success in providing ecological services (Wunder, 2007). This raises a number of issues. It may be extremely difficult to measure the direct environmental effects of a policy, possibly because the biophysical data or monitoring techniques are not available or because the environmental consequences take some time to emerge. Equally there needs to be a point of reference to measure the effectiveness of a policy: a baseline projected on a "without-project" scenario with which the environmental effects can be compared. To avoid these difficulties, environmental effectiveness is defined here in terms of forest use practices, rather than pure physical data (or environmental outcomes). Thus human behavior is used as a proxy for measurements of physical environmental change (such as soil erosion, water quality and forest resource) since it is safe to assume that changes in behavior, such as a logging ban or fire controls will have a positive effect on conserving the forest ecosystem and the eco-services it provides.

2. Economic impact. The World Bank estimates that roughly a quarter of the world's poor and 90 percent of the poorest depend substantially on forests for their livelihoods (Scherr *et al.*, 2004). In China, most forests are found in officially designated "poor counties" (Li *et al.*, 2000). Forestry plays an important role in the livelihoods of the rural population as a subsistence safety net, a source of cash income, a capital asset and a source of employment (Scherr *et al.*, 2004). Therefore, any policy relating to the forest ecosystem will have a profound impact on local livelihoods. International experiences show that PES are most effective when they aim at both achieving environmental goals and contributing to local livelihoods (Scherr *et al.*, 2006). It is therefore imperative to include economic impacts within the policy evaluation – whether or not the policy explicitly takes rural livelihoods into account. The livelihoods of local people depend on forest resources and the sustainability of the payment schemes is also determined by the intensity of the plausible negative impacts on their livelihood. Payment schemes can only

work on a long-term basis if local communities benefit from them. While some researchers mention the potential of PES to increase the income of smallholders as suppliers of forest ecosystem services, concern has also been raised about the potential adverse impacts of such schemes on rural livelihoods.

Economic impact is a broad concept which includes both the direct and indirect influence of a policy. These impacts can be interrelated and they cover different geographical and time scales. Therefore it is necessary to define what kind of economic impact we are referring to. To avoid a loss of focus, this thesis is mainly concerned with the economic impact arising directly from changes related to forest use practices. Payment schemes change the forest use practices of local communities and these new practices can have positive or negative economic impacts on the local population. The evaluation will therefore seek to establish the causal links between payment schemes, forest use practices and economic impacts. These links may be direct such as loss of income from timber sales or receipt of governmental payments. There might also be indirect links, such as small businesses that benefit from eco-tourism. In order to weigh these costs and benefits, a cost-benefit analysis has been employed to evaluate the economic impact of payment schemes on local forest users and owners.

3. Participation. Traditional environmental programs and projects which evolve as top-down, managerial policies seldom encourage, or recognize the importance of participation (Kosoy *et al.*, 2008). Since the 1980s, many development and environmental programs and projects, especially those supported by international donors, have attempted to integrate a participatory component. Researchers have analyzed the factors that drive people's participation in PES. Most of this research has been carried out in Latin American countries such as Mexico and Costa Rica, where some key drivers for local participation have been identified. These include: formal tenure titles (Grieg-Gran *et al.*, 2005), access to financial resources (Miranda *et al.*, 2003), the contribution that the PES makes to household income and the land opportunity cost (Pagiola *et al.*, 2005; Wunder, 2005) and farm household characteristics (farm size, human capital and household economic situation) (Zbinden and Lee, 2005). However, such research has generally adopted a narrow definition of participation as the willingness and capacity of forest owners and users to enter into the PES programs. This definition of participation is not suitable for China's payment schemes. In Latin America, PES is usually a voluntary transaction between buyers (governments or international donors) and ecological service providers. By contrast, the payment schemes in China are implemented by the government on a more or less compulsory basis. Therefore, participation at the entry stage does not reflect either the actual willingness of local forest owners/users to participate in such schemes or their involvement in policy implementation. To understand the participation mechanism in a broader sense, this thesis looks at participation in terms of the degree to which local farmers are involved in each stage of payment policies. For heuristic reasons, four such stages have been identified: formulation, demarcation, management and examination. Through assessing participation at each of these policy stages, it is possible to assess how much room the current design of payment schemes in China provides for involvement of local actors and how they interact with each other. This assessment of participation also examines the accountability and legitimacy of the payment schemes. Participation in the formulation and demarcation stages can be taken as an indication of the support that local stakeholders provide to the government in its role

as policy maker and implementer. The extent to which local stakeholders support and accept the payment schemes will have important influence on the performance of such schemes in both the present and the future.

PES research has shown that these three aspects are closely related (Figure 2.5). A good participation mechanism can improve local livelihoods and thereby ensure a high rate of compliance. Participation can in itself improve trust and build shared understanding among actors, which can also directly change forest use practices (Grieg-Gran *et al.*, 2005; Landell-Mills and Porras, 2002; Pagiola *et al.*, 2005; Scherr *et al.*, 2004).

EMT provides a theoretical perspective for interpreting environmental reforms in the forest sector. Payment schemes for FES can be an excellent example for demonstrating this transformation in China's context. By borrowing instruments from different approaches to evaluating environmental policy, this thesis constructs an analytical framework to evaluate the performance of the payment schemes. The result will show the extent to which environmental improvements are taking place in the forestry sector and how the implementation of the payment schemes is changing forest use practices.

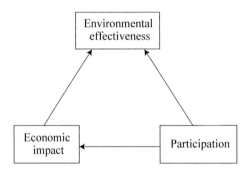

*Figure 2.5. The interrelations between the three evaluative aspects.*

# Chapter 3.
# China's forest policies – a historical transition to ecological conservation

## 3.1 Introduction

China's current forest management system has its foundation in the period of central economic planning. The current system of forest management is still characterized by national planning and control during the entire whole process, from forestry and timber production to processing and trade. Following the market-based reform that started in 1978, the forest management system changed in all aspects to adapt to the so-called socialist market economy. Simultaneously, the continuing pressure on forest resources and the related ecological crisis cried for more and a stronger control of the State on forest resources. This provides a mixed picture in China's forest sector, in which decentralization in forestry administrative system and organization parallels with centralization in public benefit forest management in the form of forestry ecological conservation projects. By the same token command and control regulation methods are complemented with market-based instruments to change forest use practices.

In this chapter, I first introduce the history of forest resources management in China from 1949 onwards. Then I describe the contemporary forestry administrative system in China and the different organizations that function together to manage forests (Section 3.2). Subsequently, Section 3.4 contributes to the main forest resources management policies that are closely related to public benefit forest in China, including forest property rights arrangements, forestland management policies, policies on classification-based management systems, and forest logging management policies. Section 3.5 introduces the major forestry ecological conservation projects in China, which play such an important role in the transition of China's forestry, from a priority on timber production to an emphasis on ecological conservation. The last section gives a brief introduction on the development of China's policy on PES.

## 3.2 A history of forest resources management in China

When the People's Republic of China (PRC) was established in 1949, the central government established a series of principles for forestry development, including protecting forest, especially in mountainous regions, promoting afforestation and forest cultivation, and reasonable use and harvesting of timber. A Forestry Ministry was established, including 4 departments (forest administration, forest use, forest management, afforestation) and a general office. The ministry initiated forest logging management policies, issued logging guidelines and formulated requirements on logging practices. Due to the demand on timber for economic development, China's forestry chose timber production as its main task. No less than 136 state-owned forest bureaus were gradually set up to harvest timber to fuel the economy in regions with large tracts of natural forests, including Heilongjiang, Jilin, and eastern Inner Mongolia in the northeast;

Yunnan, western Sichuan, and eastern Tibet in the southwest; and in parts of Xinjiang in the northwest and Hainan in the south (Xu *et al.*, 2006a).

According to the former Union of Soviet Socialist Republics (USSR) experiences on forestry development, independent afforestation and forest industry departments were set up in each forest region to counterbalance each other. However, the conflicts between these two departments made it difficult to cooperate. In the end, afforestation and forest industry departments were merged together into forestry bureaus. As a result, the forest industry branch had more power in deciding timber logging, and the afforestation branch became a subordinate in the new organization. Besides forest logging, the central government also developed forestry production plan management systems, corresponding to its economic planning system. In this system the state arranged the infrastructure in forest regions, distributed funding and equipment, made production plans for forestry bureaus, redistributed timber and forest products, and decided on the use of all forestry capital and profits (Lei, 2007). Because the forest industry worked as an enterprise, it tended to increase timber harvesting without taking into account the sustainability of forest resources. In addition, the central government formulated policies on timber trading, in order to control timber trade. In state-owned forest regions, forestry departments directly managed their timber trade. In southern collective forest regions, forest industrial bureaus were built to monopolize timber trade. In 1955, the *Temporary Arrangement on Timber Redistribution Nationwide* was carried out. All timber from China's southern collective forest regions was monopolized by the government and private timber trade was reconstructed under socialistic principles. In timber trading, governments decided on the type, amount and price of wood. This policy eliminated timber pricing mechanisms, cut off the connection between forest owners and the market, discouraged investment on forestry, and resulted in overcutting for short term benefit (Lei, 2007).

During the land reform from 1950 to 1952, the government confiscated forestland from large landlords and redistributed it to local farmers. Afterwards, the government started a collectivization movement on private land, which reorganized farmers' forest lands in the hands of village cooperative organizations. Collective forest tenure was established and has still existed as the main ownership structure in China's rural areas until recent collective forest tenure reform (CFTR).

From 1956 to 1965, China endured a range of political movements and natural disasters (Three Years of Natural Disaster, Great Leap Forward, and People's Commune Movement). These zealous political movements, especially the Making Steel and Building Canteen movement, brought a vast damage to forest resources nationwide (Shapiro, 2001). Although the *Rules of Forest Protection* was issued in 1963 to emphasize afforestation and forest protection, it had not been implemented effectively due to the political upheaval. From 1966 onwards, China entered into a ten year political upheaval, the Cultural Revolution. The whole system of forest management was broken down. State-owned forestry enterprises (SOFEs) and public facilities were decentralized to provincial or lower levels. Provincial forestry departments were closed or degraded as a subordinate of local revolutionary committees. Forest management entered into chaos and one of its previous basic principles – emphasizing on afforestation – was replaced by "food as the priority". Large forests were cleared for planting food crops, cash crops and cotton. Private forests, regarded as "the tail of capitalism", were confiscated and handed over to collectives. The Cultural Revolution movement disordered regular forestry activities and abolished forest management institutions and technical guidelines (State Forestry Administration, 1999).

After the Cultural Revolution, forest management quickly recovered following the new era of post 1978 market-based reform and opening up policy. Forest management institutions and organizations were rebuilt. A Forestry Ministry was set up in 1979 and a new department – the department of forest resource management – was established to become responsible for afforestation planning, resources survey, and statistics. In South China's collective forest areas, forest administrative organizations at county and township levels were also set up, including local forest public security, prosecutors and courts, township forestry stations, and timber check points. In 1979, the State Council issued the *Notice on Protecting Forest and Forbidding Illegal Logging*, which stipulated important items about maintaining forest ownership, forbidding illegal logging, stopping clearing forests, and strengthening timber market management. The Standing Committee of National People's Congress recognized the *Forest Law* (trial), which clarified the main principles for forestry development, including taking afforestation as a main task, and strengthening afforestation and forest management. In 1981, the Central Committee of CCP and the State Council issued the *Decision on Several Issues on Protecting Forest and Developing Forestry*, which put forward that "forest ownership should be stable, *ziliushan* (private forest plots) should be demarcated clearly, and forestry production responsibility system should be established" (Central Committee of Chinese Communist Party and Chinese State Council, 1981). Subsequently, each province issued forest tenure certificates and distributed collective forests to farmer households as *ziliushan* or *zerenshan* (collective forest plots under private management). The new institutional arrangement on forest property rights encouraged farmers to invest in forestry. Furthermore, in 1985, the Central Committee issued *Ten Policies on further Promoting Rural Economy*, which abolished the governmental monopoly on timber trade and allowed free trade of private or collective timber on the market. However, further loosening the control on forest resources intrigued overcutting and illegal logging nationwide. So in 1987, the Central Committee issued the *Order on Strengthening Forest Resources Management in South China's Collective Forest Areas and Prohibiting Illegal and Over-cutting*, which stipulated that for major counties with timber production, forestry departments continued their monopoly on timber trade in order to prevent illegal and overcutting. After the Order was implemented, provincial governments announced local forest laws and regulations to regulate forest management. The local forest management law system was formed in this stage (State Forestry Administration, 1999).

In 1985, a new Forest Law was published, which strictly controlled forest logging and stipulated that consumption on timber forests should be lower than natural growth. In the same year, the Forestry Ministry issued *Temporary Regulation on Making Forest Logging Quota* and it regulated detailed procedures for implementing logging quotas. In 1987, the State Council forwarded the Forest Ministry's *Examination Report on Annual Forest Logging Quota for Each Province*. Its implementation embodied the principle of sustainable development in forestry and meant that China's forest management transformed from timber production plan to forest logging quota.

In the same period, the central government also started to use ecological projects to promote reforestation. Since the start of the Shelterbelt Program in Three-North Area in 1978, the government initiated 17 forestry-related ecological conservation projects, including the Shelterbelt Program along the Upper and Middle Reaches of Yangtze River, Coastal Shelterbelt Program, Sandification Control Program, Greening Taihang Mountain Program, Greening Plain Program and so on (Zhou, 2001). In addition, 30 million RMB were invested annually by the State for developing timber plantation bases. Local governments also promoted ecological conservation initiatives. For example, Guangdong Province made a decision on speeding afforestation and greening the entire province. After 1985, 12 provinces made similar commitments and completed the greening of barren mountains within their jurisdiction (Lei, 2007). In 1990, the State Council issued the *Afforestation and Greening Outline from 1989 to 2000*, which clarified guiding principles, objectives, planning, tasks and priorities in afforestation nationwide and promoted afforestation and greening of barren mountains in provinces. However, in this period, timber production was still the core business of forestry development in China. The management system, institutions and ideas from planned economy period were still overwhelming in China's forest management.

From 1998 onwards, China's forest management entered into a new phase: a historical transformation from timber production to ecological conservation (Zhou, 2002). The 15th Conference of the Central Committee of CCP took afforestation, soil erosion control and anti-desertification as important national strategies facing the new century. In 1999, the State Council issued the *National Ecological and Environmental Development Outline*, which stressed that ecological needs became the priority of forestry development. Two important policies on forest management and protection were initiated. The first was the six forestry ecological conservation projects, which invested tremendous funds into forest protection and ecosystem restoration (see Section 3.5). The other was forest ecological compensation policy, by which the principle of PES was built into China's forest management.

## 3.3 China's forestry administrative system and its organizations

China's forest administration system has five levels, along its governmental tiers (see Figure 3.1) (Li *et al.*, 2006a). At the national level, there is the State Forestry Administration (SFA), which is responsible for providing guidance to the forestry sector and for supervising forest management and protection over the entire nation. Each provincial forestry department conducts forest management on state-owned forestland within its jurisdiction by state-owned forestry enterprises, bureaus or farms; and they take charge of supervising forest protection on collective forestland. Similarly, municipal and county forestry bureaus also take care of their forests within their jurisdiction. Depending on their administrative level, state-owned forestry enterprises, industrial bureaus and farms are generally subject to governments at different levels. Sometimes, they have the same position as forestry bureaus in the governments, but at other occasions they are lower in hierarchy than the local forestry bureaus and function under their management. Township forestry stations, as grass-root forestry management organizations, directly organize forest management and protection at township level. Out of this five-tiered administrative system, four big state-owned forest industrial groups, which manage large tracts of state-owned forest in northeast China, are directly under the State Forestry Administration[7]. These four groups are separate from the tiers, because historically they have been established in the period of planned economy as main timber producers and the state required more direct control on such state-owned enterprises to complete its timber production plan. Currently, the relationship between these

---

[7] Actually, there is an on-going decentralization process among these four state-owned industrial groups. The management rights are transferred from SFA to provincial governments.

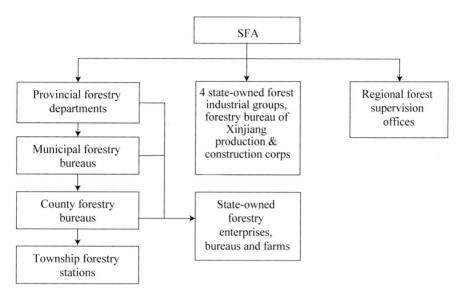

*Figure 3.1. China's forest administrative system.*

forestry industrial groups and SFA becomes more and more loose. The provincial governments are gaining more control on these industrial groups, especially on issues like appointment of staff. Under the existing administrative system, provincial and county forestry bureaus function as a component of local governments and financially rely on local governments. SFA can only sustain a loose control on the behavior of local forestry authorities. Therefore, SFA established 14 regional forest supervision offices, which are responsible to directly supervise forest management and protection of several provinces within the regions.

This complicated management structure brings problems to forest management and lowers the efficiency of implementation. In China's multilevel political system, the provincial governments have the same bureaucratic rank as ministries. In 1998, the Ministry of Forestry was downgraded to a bureau-level organ – the State Forestry Administration. SFA became lower in bureaucratic rank than provincial governments. It only can provide sectoral guidance for provincial forestry departments. Provincial governments have more influence on the decision making of local forestry departments. In some cases, the provincial governments prefer economic growth to forest protection and some nature reserves and public benefit forest are sacrificed to provide land for development. Second, forestry industrial groups and state-owned forest enterprises carry out daily forest management, planning and monitoring by themselves. They usually have the same bureaucratic rank as local forestry departments or bureaus. Therefore, local forestry authorities cannot supervise their practices in forest management. The income of forest enterprises largely depends on logging and they tend to harvest more forest than logging quota allows to support livelihood of their employees[8]. Furthermore, state-owned forest enterprises and farms, which have

---

[8] These forest enterprises usually hired too many employees in the period of planning economy and cannot downsize their employment by themselves.

higher bureaucratic ranks than local governments, often have conflicts with local governments on issues including land administration, forest management and ecological conservation.

## 3.4 China's forest resources management policies

China's forest resources management policies are based on the *1984 Forest Law* (revised in 1998), but are still constantly changing to adapt to the developing market-based economy and to deal with the emerging ecological crisis in forests. The forest management policies construct a fundamental institutional setting for payment schemes and influence effectiveness and efficiency of implementation of the payment schemes. In this section, the main forest resource management policies are introduced, as well as the detailed procedures and measurements for China's forest management, distinguishing between commercial forests and public benefit forests.

### *3.4.1 Forest property right arrangements*

China's forest property rights are divided into two legal objects: forestland and forest timber. All forestlands are owned by the State or collective economic organizations, usually villages. Collectives control approximately 60% of China's total forestland and the other 40%, primarily with natural forests, is managed by the State in the form of 3,000 independent state-owned forest farms and 136 state forest bureaus[9] (Li *et al.*, 2006a). State-owned units, such as state-owned forest farms and state forest bureaus, can use state-owned forestland but do not have ownership. Collective economic organizations, private companies and individuals can use collective or state-owned forestland through responsibility contracts, rental contracts and transferring contracts. By such contracts[10], private companies and individuals can use the forestland for forestry activities within a given period, which is usually shorter than legally-binding terms for forestland renting (30 years for collective forestland) but the renters cannot change the land use towards for other uses. As a result, the ownership and use rights of forestland often do not fall on the same entities. There is, however, one exception. Historically in the 1980s, villages have distributed some collective forestlands to individual farmer households, which can be inherited within the village but cannot be transferred to others outside the villages. Under such circumstances, these forestlands, so called *ziliushan,* are de facto owned by individual farmer households with some limitations, although legally it is still a collective property. Comparatively, forest timber has a relatively uniform structure of property rights. Forest timber can be owned by the State, collective economic organizations, farmer households, private companies and individuals. Following the market-oriented reform, forest owners become more and more diverse and the fragmentation and diversification of forest ownership is creating tremendous challenges for Chinese forest management. Compared to the old situation under central economic planning, today forest authorities have to face numerous forest owners and have to deal with more complex relationships among different owners, such forestland

---

[9] The 136 state forest bureaus have been set up during the era of planning economy. They are line managed by local or provincial governments, but are subject to the SFA guidelines on forestry practices.

[10] For collective forestland, the members of the collective have the priority to rent the land for forestry activities.

owners and forest timber owners. Public benefit forest management also has been impacted by such changes. In some regions, such as Liaoning, Jiangxi and Guangxi, public benefit forest is redistributed from villages to individual farmer households. Under these conditions existing payment schemes for environmental services are questioned regarding their capacity to provide enough compensation to individual farmer households (State Forestry Administration, 2006a). The governance structure behind these payment schemes, which relies on command and control measures to directly supervise villages, seems incapable to make sure the public forest plots – in the hands of more and more fragmented small forest owners – are used in a safe and sustainable way.

### 3.4.2 Forestland management policies

Nowadays, the Chinese government implements a set of strict forestland management policies following a vast loss of forestland from rapid development of construction and mining. Over the past decade more than 6 million hectare of forestland was cleared annually for construction and mining (Zhang and Wang, 2000). Current forestland management policies stipulate that all occupation and acquisition of forestland should be approved by forestry departments; if more than 35 hectare of ordinary forestland or more than 10 hectare of public benefit forestland is occupied for other purposes, it has to be approved by the national SFA (Li *et al.*, 2006a). This approval policy plays an important role in preventing public benefit forest being changed into construction or mining sites during the current period of rapid economic growth. Furthermore, the policies also require the units that occupy forestland for development to provide plantation restoration funds for local forestry departments, which than can plant trees at other sites in order to keep the total area of forestland stable. In addition, the Forest Law stipulates that no forestland can be converted into cropland and the use rights of public benefit forest are not allowed to be transferred on the market[11], in order to protect forest resources and reduce the risk of deforestation on public benefit forest.

### 3.4.3 Classification-based management

Classification-based forest management is a fundamental policy in China's forest sector. This idea first appeared in the document, issued by the Central Committee of CCP and the State Council in 1981, entitled *Decision on several issues on forest protection and forestry development* (Central Committee of Chinese Communist Party and Chinese State Council, 1981). This document requires that each province should designate different forest types to its forestland and make different management plans for 5 forest types: timber forest, protection forest, economic forest, firewood forest, and special use forest. In 1995, classification-based forest management was listed as a management reform strategy in the *Outline for forestry economic system reform,* issued by the former State System Reform Commission and the former Ministry of Forestry (State System Reform Commission and Ministry of Forestry, 1995). Sixty counties in 15 provinces or autonomous

---

[11] This policy has been implemented less strictly in some provinces due to the collective forest tenure reform, where there is a pilot that public benefit forest is allowed to be contracted out to other individuals or companies.

regions were selected for experimenting classification-based forest management and a leading group in each county, which included forestry bureaus, agricultural bureaus, financial bureaus, reform commissions and other departments, was established to direct the reform on classification-based forest management in 1996. However, the progress of the reform was slow due to insufficient financial support by the central level. The reform did not really obtain momentum until the central payment scheme was established in 2001. Each province set up its own leading group of counties for the reform. Especially, 8 provincial governments (including Yunnan, Guangdong, Hunan, Fujian, Liaoning, Heilongjiang, Jilin, Henan) directly participated in planning and organizing the reform and showed strong political willingness to apply the classification-based forest management.

The objective of the reform was to establish two types of forest management systems, one for public benefit forest and one for commercial forest. All forest resources are divided into these two types of forests and different policies concerning financial support, use rights transfer, logging, acquisition, and administration rights are applied to them. At present, the reform is still underway in several provinces and especially the provinces involved in the Natural Forest Protection Program (NFPP) just started to reclassify their natural forests under protection into public benefit forest in 2010.

### 3.4.4 Forest logging management

China's forest logging management system includes 3 main components: forest logging quota management, timber production planning, and a logging license system. Forests have been overharvested widely during the period of China's centrally planned economy up till 1978[12] (State Forestry Administration, 1999). In order to reduce overcutting, China's government started to implement logging management policies and practices in 1979. The SFA started to work out logging plans for each county or state-owned forest bureau, in line with the Forest Law's principle that "consumption of forest should not be higher than its growth". However, this policy, strongly in line with a central planning economy, did not function well and the economic reform during the same period actually induced more severe – and especially illegal – cutting. In 1987, the logging plans were revised into annual logging quotas and complemented with annual timber production plans, in order to more strictly control forest logging. SFA was responsible for making a national timber production plan and national forest logging quota, and distributed planned timber harvests and permitted quota to each county via the provincial levels or state-owned forest bureau. All harvests in each county or state-owned forest bureau, both those planned and those outside of the plan, should be lower than the annual logging quota allowed for them. At the same time, a logging license system was established, which requires that any individual or collective organization has to apply a logging license from local forestry authorities (usually county forestry bureaus) before forests are harvested.

Forest logging management differentiates between commercial forest and public benefit forest. Clear cutting can be operated only for commercial forests, and in public benefit forest only strict selective cutting is allowed, which requires relatively higher costs and more labor input than clear

---

[12] Actually, after more than ten years' limitation on forest harvesting, in state-owned forest region in northeast China, overcutting is still practiced by local forestry bureaus due to the need for local livelihood.

cutting. In practice, local forestry authorities hesitate to issue logging licenses for public benefit forests because of the high risk of illegal logging and high cost of monitoring.

Forest logging quota management makes a contribution to forest protection nationwide, but it is not in line with timber production and trade in a market-based economy, in which forest owners have to consider timber prices on the market in their logging decisions. In addition, the central government anticipated that harvest limits could maintain or improve forest growth rates. However, monitoring and enforcing local manager decisions is difficult under current forestry administrative system, and state forest managers lack incentives to harvest and reforest according to the quotas. Xu and colleagues (2003) found that higher quotas led to decline in forest growth over time due to undetected overcutting or under-reforestation by state forest managers. Lately, logging management is undergoing a transition from using forest quota management to control timber harvesting, to using forest management plans to direct timber production. It is designed to provide a flexible choice to forest owners to encourage them to increase input in forests and at the same time to sustain forest growth for maintaining ecological objectives.

## 3.5 China forestry ecological conservation projects

Ecological conservation projects have been developed after 1998 and have turned into the main conveys of forest protection activities in China since then. They rely on significant financial support from the central government and have been implemented along the administrative line. There are six major ecological conservation projects, which signified a turning point of China's forestry development. Traditional forestry focused on timber production and all investment, measures, and policies were developed around this focus. However, the ecological conservation projects after 1998 set a new backdrop which put ecological conservation as the priority of China forestry. Not only are a lot of policies developed to create institutional support to divert forestry practices, tremendous central investment is mobilized to provide a solid foundation to develop these projects.

### 3.5.1 Conversion of Cropland into Forest and Grassland Program

Conversion of Cropland into Forest and Grassland Program (CCFGP) as the largest ecological restoration project in China covers 2,279 counties in 25 provinces (autonomous regions and municipalities) and targets the ecological sensitive regions in western China, including sloping cropland along rivers, lakes and reservoirs, and cropland within the areas stricken by soil erosion and sandstorm (Figure 3.2; State Forestry Administration, 2009a). It aims to tackle soil erosion in ecologically fragile areas by returning 14.66 million hectares of croplands to forests and afforesting 17.33 million hectares of barren hills and wasteland from 2001 to 2010. Under this program, farmers can choose to convert their sloping land into ecological or economic forests, but the ecological forests should account for over 80% of the total converted land within each county (Chinese State Council, 2003b). The government compensates the farmers with different types of subsidies for 5 (for economic forests) or 8 (for ecological forests) years, which includes grain (1,500 kg per ha each year within the Yellow River Basin and 2,250 kg per ha each year within the Yangtze River Basin), cash (3,000 Yuan per ha each year) and free saplings (Zhou, 2001). However, in some regions, the grain subsidy is replaced by cash, which is calculated in a fixed price (1.4

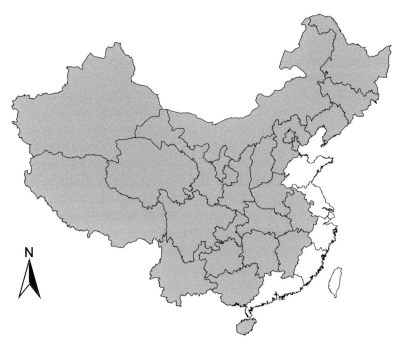

*Figure 3.2. Provinces implementing Conversion of Cropland into Forest and Grassland Program (grey).*

Yuan per kg), due to the significant draw-down of State Grain Bureau stocks for program subsidies and the later rise of grain prices (Bennett, 2008). After the first phase of the project, the central government continues to provided subsidy to farmer participants during the extension of the program (5 years for economic forests and 8 years for ecological forests), but the grain subsidy is adjusted to only cash (1,575 Yuan per ha each year within the Yangtze River Basin and 1,050 Yuan per ha each year within the Yellow River Basin). The previous cash subsidy of 3,000 Yuan per ha still is paid to farmers on the condition that they continue to manage the young forests (State Forestry Administration, 2009a). Besides the subsidy for farmers, the central government also arranged some fiscal subsidies for local governments to compensate their loss of local fiscal revenue.

### 3.5.2 Natural Forest Protection Program

The Natural Forest Protection Program (NFPP) covers 734 counties and 167 state forest bureaus in 17 provinces (autonomous regions and municipalities) (Figure 3.3; Zhou, 2001). It has been established by SFA in 1998 and implemented in China's key state-owned forest areas, including the upper reaches of the Yangtze River, the upper-middle reaches of the Yellow River, northeast China and Inner Mongolia. The program aims to rehabilitate and protect natural forests by transforming forest production practices in these areas. First, commercial logging is banned in natural forests in the upper reaches of the Yangtze River and the upper-middle reaches of the Yellow River. The timber output in the program area was planned to be reduced by 19.91 million cubic meters from 1998 to 2010 (Zhou, 2001) and has been reduced by 19.98 million cubic meters till 2010 (Zhao,

*Figure 3.3. Provinces implementing Natural Forest Protection Program (grey).*

2011). Although the logging is still higher than the growth of the forests in the regions, it proved really a difficult process of implementing these policies in state-owned forest areas, especially in northeast China and Inner Mongolia, as most local revenues were generated from forests and local livelihood heavily depended on forests. Second, an additional 10 million hectares of forests was established in the program area and the forest cover rose by 3.7 percent from 1998 to 2010 (Zhao, 2011). In addition, the program also shut down small timber processing companies and supported laid-off lumberjacks and workers to find alternative jobs. About 741,000 redundant forest workers in the program areas were diverted and re-employed in other sectors up till 2010. Although the program achieved successes, especially in protecting natural forest, the quality of forests is still quite low; the regions are still under the threat of ecological crises such as soil erosion and sandification; and the local economy is still underdeveloped and could easily revert to rely on timber harvesting in future (Zhao, 2011).

In 2010, the central government decided to extend the program and a second phase of implementation started, to run from 2011 to 2020. New agendas, like maintaining national timber security and combating global climate change, have been written into the planning and program implementation. Eleven counties along the catchment of Danjiangkou Reservoir were added into the program. For northeast China and Inner Mongolia the aim is to establish both a national strategic reserving base for timber production and an eco-safe barrier for north China. The plan for this region includes reducing annual timber production from 10.94 million cubic meters to 4.03 million cubic meters from 2011 to 2015; increasing the forest area by 600 thousand hectare, the stocking volume by 290 million cubic meters, and the forest carbon sink by 109 million tons; and providing 443.2 thousand jobs in forested regions up till 2020 (State Forestry Administration, 2010b). In the upper reaches of the Yangtze River and the upper-middle reaches of the Yellow River, the program aims at building a stable forest ecological barrier for the region. The objectives include continuing stopping commercial logging on natural forest; increasing the forest area by 4.6 million hectare, the stocking volume by 810 million cubic meters, and the forest carbon sink by 307 million tons; and providing 205.3 thousand jobs in forested regions till 2020 (State Forestry Administration, 2010c).

### 3.5.3 Other forestry ecological conservation projects

Besides NFPP and CCGFP, there are 4 relatively small conservation projects. The first one is the Wildlife Conservation & Nature Reserve Development Program (WCNRDP). It aims to improve nature reserves and strengthen the protection of species, and wetlands. The second one is the Shelterbelt Program (SBP), which covers all 31 provinces. The program is designed to establish new shelterbelts and improve low-efficient shelterbelts to control desertification in the three-north regions (North, Northeast and Northwest China) and ameliorate ecological crisis in other areas. The third one is a regional project, Sandification Control Program in the Vicinity of Beijing and Tianjin (SCPVBT), which covers 75 counties in Beijing, Tianjin, Hebei, Shanxi and Inner Mongolia with a combined area of 460,000 square km. By returning farmland into forest and controlling overstocking on grasslands, the project aims to lift the hazard of sandstorms in areas surrounding the capital city. It was also an important program to improve the Beijing environment to meet the needs of the 2008 Olympic Games in Beijing. The last one is Fast-Growing and Fast-growing and High-yielding Plantation Development Program (FHPDP). It is a program to support other conservation programs by constructing commercial timber bases to ease the shortage of timber supply and reducing the pressure of timber demands on protected forests. The program covers 114 state forest bureaus or farms and 886 counties in 18 provinces (autonomous regions), which are located east of China's 400 mm rainfall line (Zhou, 2001).

## 3.6 China's policy on payment for ecosystem services

With a rapid growth of the economy, ecological degradation and environmental pollution have become bottlenecks for social and economic development in China. Facing serious ecological crises, the Chinese government pursues the establishment of an integrated ecological compensation mechanism to gradually amend and partly replace traditional administrative measurements solely relying on command-and-control, and to provide institutional support to the transition to a resource-conserving, environment-friendly society.

### 3.6.1 General background on China's payment for ecological services

Since 2000, the Chinese government has set up a series of policies, laws and regulations and provided finances to support ecological compensation mechanisms. Financial compensation has been included into governmental agendas at all levels as an important element in fulfilling their tasks on environmental protection. In 2005, the State Council issued the *Decision on Implementing*

*Scientific Concept of Development and Strengthening Environmental Protection*, which "requests the central and local governments to improve payment policies for environmental services and to establish eco-compensation mechanism as soon as possible, ask to take eco-compensation into account while the central government provides fiscal support for local governments, and encourage piloting eco-compensation schemes at national and local levels" (Chinese State Council, 2005). In 2010, the State Council listed *Regulation on Eco-Compensation* within its legislation plan and the State Development and Reform Commission (SDRC) led a drafting group with more than ten governmental departments, including the Ministry of Finance (MoF), the Ministry of Land and Resources, the Ministry of Environmental Protection (MEP), the Ministry of Agriculture, and the State Forestry Administration (SFA). Currently (November 2011), the legislation on the regulation is still in draft form. Although formal regulation is not in place, China has initially set up a framework on eco-compensation at different levels, covering various types of ecological services. Existing eco-compensation schemes include trans-regional schemes, payment schemes for ecosystem services (covering forest ecosystem, grassland ecosystem, wetland ecosystem, and nature reserves), payment schemes for important ecological functions (including water source conservation, biodiversity, windbreak and sand fixation, soil erosion prevention, and flood control), and payment schemes for exploiting natural resources.

### 3.6.2 China's payment policies for forest ecological services

As an important part of China's eco-compensation mechanism, payment policies in the forest sector started earlier than other sectors. With tremendous investment from the central government, the payment policies for forest ecological services have made great achievements and gained precious experiences for applying PES in other sectors. As early as 1992, the State Council issued *Notice on Important Points of Economic System Reform in 1992*, which specified that governments should "establish forest pricing mechanism and forest eco-compensation institution to impose cost on forest ecological services" (Chinese State Council, 1992). Article 8.6 of the Forest Law (revised in 1998) stipulated that "the State establishes the forest ecological benefit compensation fund to be used for the planting, tending, protection and management of the forest resources and woods for shelter forests and special-purpose forests either of which generate ecological benefit" (NPC, 1998). Following the principle of PES promoted by the Forest Law and other official documents, the central government initiated a series of forestry ecological conservation programs, including NFPP and CCFGP. These programs included subsidies to encourage a change to sustainable forest use and ecological conservation practices, and as such replace traditional methods solely relying on command-and-control. Hence, a new institution is established to govern how forest resources should be used, and what forest owners can get for providing ecological services. Furthermore, in 2001 the central government piloted a forest ecological benefit compensation fund. Three years later, it formally set up the fund and the MoF and the SFA issued *Measure on Management of the Central Forest Ecological Benefit Compensation Fund* (Ministry of Finance and State Forestry Administration, 2004). It meant that China established a formal forest eco-compensation institution and the principle of PES was gradually applied in the forest sector for ecological conservation. Recently, Chinese forest eco-compensation institution has been improved with respect to its coverage, investment and payment standards. Up till 2010, it has covered 70 million hectare of forests, accounting for 22.88% of the total forestland nationwide; the total annual payment has reached 10.18 billion RMB; and the payment standard for collective and private forests has been doubled from 75 RMB per ha to 150 RMB per ha per year (State Forestry Administration, 2011). In 2009, the SFA and the MoF together issued *Measures on Demarcation of National Public Benefit Forest*, which stipulated that new plantation of forests within the CCFGP, and the forests of new nature reserves and along new reservoirs can be covered by the compensation fund and receive subsidy from the central government for forest management. It also allowed private forests (transferred from the villages to individual farmer households during the forest tenure reform) to quit from the central payment scheme, as long as the forests have no major impact on the local environment (State Forestry Administration, 2010a). In addition, after 2010, CCFGP and NFPP entered into their second phase of implementation. There is a trend for including forests protected by NFPP and created by CCFGP into the central and provincial payment schemes for public benefit forest. In the process of China's ecological modernization in the forest sector, conservation projects serves as a policy punch to change local forest use practices over a large area in a short period. The payment schemes for public benefit forest play a role in providing a stable mechanism to sustain efficient management on forests under these ecological conservation programs.

# Intermezzo: research methodology

## I.1 Introduction

As the main research questions – formulated in Chapter 1 – illustrate, the objectives of this research are to assess the ecological and socio-economic effects of forest PES schemes in China, to assess participation of state and non-state actors in the implementation of the schemes, and to analyze the influence of forest tenure reform on the functioning and outcome of the schemes. A research approach combining both qualitative and quantitative methods is used to achieve these objectives. It is reasonable to use quantitative methods to assess the effects of the schemes on the environment and on the socio-economic circumstances of farmers. At the same time, qualitative methods are required to better understand how the schemes function in a specific local context, how farmers and other agencies participate in the design, implementation and assessment of the schemes, and how the characteristics of the schemes are (causally) related with the performance of the schemes. Before proceeding to the empirical chapters, this intermezzo addresses a number of methodological concerns. In Section I.2, the research design is introduced in detail. Section I.3 provides the methods for data collection used in this research.

## I.2 Research design

A research design not only has to conceptualize an operational plan to undertake tasks and procedures to answer research questions, but also to ensure that the answers generated from the procedures are valid, objective and accurate (Kumar, 2005). There are various research strategies that can be chosen, such as case studies, experiments, surveys, and analysis of archival records. Choosing a research strategy depends on the type of research questions posed, the extent of investigator's control over actual behavioral events, and the degree of focus on contemporary as opposed to historical events (Yin, 1994). According to the research questions and objectives, this research covers the implementation of major payment policies for forest ecological services in China, evaluates their performance, and assesses how structural changes in the context impact on the design, implementation and performance of these schemes. In general, it aims at answering so-called "what" questions – the outcome of the payment schemes in different aspects. Obtaining such outcomes is more likely to favor a survey strategy. However, the payment schemes have been implemented across the country and it is impossible to carry out a nation-wide survey with the limited time and resources given for this research. Also, some of the research questions are typical "how" question, for which a survey research design will not be adequate. In addition, it is impossible to use an experimental design to rule out the impacts of external factors, since the payment schemes are usually implemented at the same time within a whole province or region. For these reasons, case studies are employed as a convenient and practical way to narrow down the study sites, without too much reducing validity and representativeness. To rule out to some extent the impacts of other external factors, as much as possible representative cases have been selected. Next, I introduce why the Forest Ecological Benefit Compensation Fund Program (FEBCFP) is selected as the typical payment schemes and how within this program three case study provinces (within each province about 5 villages were selected as case study site) are chosen for this research.

As introduced in Chapter 1 various major forest-related policies implemented in China can be identified as public fiscal payment policies for FES, including Natural Forest Protection Program (NFPP), Conversion of Cropland to Forest and Grassland Program (CCFGP), Forest Ecological Benefit Compensation Fund Program (FEBCFP), Sandification Control Program for Areas in the Vicinity of Beijing and Tianjin (SCPVBT), Shelterbelt Program in Three North area and along Yantze River (SBP), and Wildlife Conservation and Nature Reserve Development Program (WCNRDP) (Table I.1). Among them, NFPP, SBP and WCNRDP have stronger administrative/central planning features but weak economic incentives, while CCFGP, FEBCFP and SCPVBT have a much stronger emphasis on economic incentives (such as cash or grain subsidies) aimed at encouraging provision of environmental services (Sun and Chen, 2006).

Owing to restrictions in time and resources, it was not possible to include all these payment policies in the research design. Hence, the most representative and at the same time promising payment policy was include in our in-depth case study. Since China is in an economic transition from central planning to a market-based economy, the administrative type of government-run programs, such as SBP, are becoming increasingly inappropriate (Sun and Chen, 2006). Furthermore, policies with economic incentives have the character of compensation for the loss of forest owners, which is caused by the policy restriction on the utilization of their forest or land, while other policies mainly resort to regulation to limit logging forest. This research will focus on policies adopting economic incentives as the main mechanism to obtain sustainable forests. Within this category, most attention from academia has focused on CCFGP and there is now a wide (evaluation) literature on this program. The SCPVBT is a regional project, supported by central government, which has been carried out in vicinity of Beijing and Tianjin to control sandification, with policy measures similar to CCFGP: the use of cash and grain subsidies and free saplings provision to motivate farmers to converse their cropland into forest or grassland. FEBCFP has hardly received scholarly attention, also because the investment is relatively small compared to some of the other ecological programs. FEBCFP has been formally written into the Forest Law and gained a stable financial avenue in the budget of the central government, without duration limit. Furthermore, the trend that FEBCFP increasingly covers forestland replanted by ecological

*Table I.1. Major public fiscal payment policies in China (updated with data from Zhou, 2001).*

| Payment policies | Attribute | Coverage (#provinces) | Investment (Yuan) | Implementation (year) | Main policy measures |
|---|---|---|---|---|---|
| NFPP | Program | 17 | 96.2 billion | 10 | Administrative |
| CCFGP | Program | 30 | 358.4 billion | 10 | Economic incentives |
| FEBCFP | Fund | 31 | 3 billion annum | unlimited | Economic incentives |
| SCPVBT | Program | 5 | 62.8 billion | 10 | Economic incentives |
| SBP | Program | 31 | 57.7 billion | 10 | Administrative |
| WCNRDP | Program | 2,500 nature reserves | 135.6 billion | 30 | Administrative |

programs such as NFPP, promises that in the end forests crucial for national ecological security would be included into FEBCFP. In addition, more than 25 provinces have followed the central FEBCFP by launching their own FEBCFP-like payment schemes. Rather than a one-shot project with funding solely from the central government like the other five programs, FEBCFP serves as a lasting institution supported by the Forest Law and prompts an array of local initiatives within its policy framework. FEBCFP and its local counterparts signify a more institutionalized solution to maintaining forest ecological services in China and do further away with traditional command and control measures than other Chinese forest protection programs. Accordingly, FEBCFP and these local payment schemes are selected as public payment policies for this research.

Three case provinces – Fujian, Guangxi and Liaoning – have been selected out of the 31 provinces working with this program (see Figure I.1). There are several reasons for selecting these three provinces (or regions) as cases. First, they have been included as the first batch of pilot regions into the central FEBCFP since 2001. This means that the payment schemes have been implemented in the regions for more or less a decade. A long history with the payment schemes assures a higher plausibly of more profound impacts of this scheme on local forest use practice. Second, the selection sufficiently represents the geographical variation of program areas, in terms of forest coverage, economic and social development. The three cases are located in the north and south parts of China, include rich (Fujian) and poor (Guangxi) provinces, areas with high

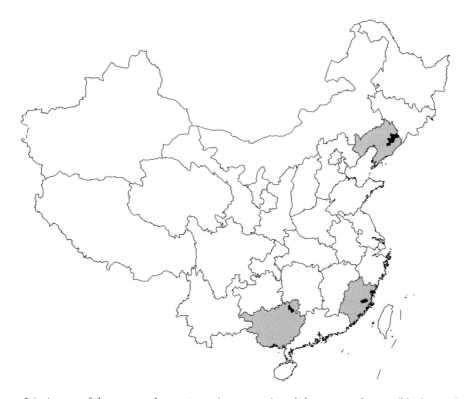

*Figure I.1. A map of the case study provinces (grey areas) and the case study sites (black areas).*

forest coverage and areas with less forest coverage, and areas where forestry still is a major sector of the economy (Liaoning) and areas where that is not so much the case. Third, this selection can exclude the impacts of other ecological conservation programs such as NFPP and SCPVBT, as these have not been implemented (yet) in these three provinces. Forth, forest tenure reform, as an important institutional factor interfering with the incentive schemes, is changing the structure of forest property rights in China. This institutional transition interacts with the implementation of payment schemes and its impact has to be considered during the policy evaluation. The three selected provinces show variations in the implementation of the forest tenure reform in public benefit forest. These differences can be built into the research design to explore the impact of the reform on the payment schemes through case comparison.

In each province, the case studies for evaluating payment schemes are conducted within two counties and 2-5 administrative villages, selected according to the following criteria:

The scale of program investment. Priority will be given to a case study site where there is continuous and large investment for forest ecosystem services.

Natural and environmental conditions. Superior or inferior natural and environmental conditions have a close bearing on the necessity and effects of payment policies. Therefore, natural and environmental conditions and their variation were applied as a vital factor in selecting case study sites.

Geographical location. The selection of case study sites considered geographical locations, and tried to cover a diversity of terrains and economic conditions, so as to ensure the case study was typical.

In each village, 10-15 farmer households have been randomly sampled with the help of village heads. One adult member from each household (usually the head of the household) responded to a questionnaire, which asked the farmer about household income and livelihood, forest resources, participation in payment schemes and forest tenure reform, attitudes towards the schemes and the reform, and willingness for compensation. Table I.2 demonstrates the actual sampling in this research.

## I.3 Methods of data collection

This research employs three methods for collecting data: secondary data collection, in-depth interviews with key informants and surveys (Table I.3). Through these methods, information and data about key policies and their implementation have been collected at different administrative levels and among different policy/governance actors. The information on objectives and

*Table I.2. Selected cases and sampling.*

| Cases | Counties | Villages | Farmer households |
|---|---|---|---|
| Fujian | 1 | 5 | 50 |
| Guangxi | 1 plus 3 farms | 2 | 19 |
| Liaoning | 2 | 4 | 54 |

*Table I.3. Data collection methods.*

| Data collection methods | Tools | Data sources |
|---|---|---|
| Secondary data collection | personal contacts, governmental database and websites | policy documents, statistical data, law and regulations at national level, provincial level and county level |
| | | scientific data and information in local environmental monitoring institutions |
| In-depth interviews | unstructured | nation level: departments in State Forestry Administration |
| | | provincial level: officials in provincial forestry bureaus |
| | semi-structured, face-to-face | county level: officials in county-level forestry bureaus |
| | structured, face-to-face | village level: heads of villages |
| Survey for payment schemes | questionnaires | 3 state-owned forestry farms |
| | | 10-15 farmer households for each village |

implementation measures (inputs and outputs) of payment policies is obtained from official document and data analysis, and from face to face and telephone interviews with relevant officials. Farmer surveys have been carried out to gather quantitative and qualitative data about outcomes and impacts of, and participation in, payment policies. In addition, depending on availability, scientific data from local research institutions and monitoring stations have been collected with respect to soil erosion and forest ecosystems, as these can provide more concrete information about impacts of payment policies. The methods of data collection and possible data sources are demonstrated in detail below. A list of the interviewed respondents (with the exception of those involved in the survey) has been included in Appendix A.

# Chapter 4.
# Emergence of local payment schemes for forest ecological services in Fujian Province[13]

## 4.1 Introduction

Fujian province, in the southeast of China, consists of 9 municipalities (Fuzhou, Xiamen, Quanzhou, Zhangzhou, Putian, Longyan, Sanming, Nanping, Ningde), 14 cities at county level and 72 counties and districts, with a population of 35 million (Figure 4.1). The mountainous area in the west makes Fujian geographically isolated from other provinces. With 7.37 million hectares its forest coverage is over 60%.

Fujian Province is a relatively isolated geographical area. It faces Taiwan Strait in the east and is surrounded by high mountains (about 1000 meters high above sea level), which separate it from Zhejiang, Jiangxi and Guangdong provinces respectively in the north, west and south. This relatively isolated geographical location and its river system form an integrated ecosystem in Fujian. The upstream and downstream areas are closely interrelated. The environmental condition in downstream areas is highly dependent on the protection and management of forest ecosystem

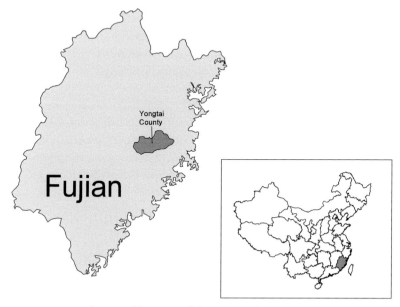

*Figure 4.1. Fujian Province in China and location of the case study.*

---

[13] This chapter is based on an article submitted to Forest Policy and Economics in May, 2010, as Dan Liang, Arthur P. Mol, Jan van Tatenhove and Yonglong Lu. Evaluating payment schemes for forest ecological services in China: the case of Fujian Province (under review).

in upstream areas (Zeng, 2003). Although it is one of the regions frequently stricken by typhoons, heavy rain, and other natural disasters, the soil erosion in Fujian is not very severe (in 2000, its land affected by soil erosion was 10.72%, lower than the average of the whole country), which is due to its high forest coverage and a vast natural forest. However, the risk of soil erosion is relatively high due to thin soil layers (average soil thickness is 20-80 cm), which is susceptible to erosion from rainfall. From 1987 to 1997, the area of natural forest in Fujian decreased from 42,570 km$^2$ to 37,030 km$^2$ (Zeng, 2003). Clearing natural forest for planting fruit trees, overcutting natural forest and mining are the main reasons for the reduction of natural forest. Following the reduction of natural forest, Fujian's biodiversity is also under threat. Since 1990s, emerging market demand for broadleaf timber and fungi cultivation has driven a vast logging on natural broadleaf forest. Only several national and provincial protected areas survived as isolated islands which hardly provide sufficient habitats for wildlife. From 2001 onwards, several payment schemes were consecutively initiated by central and provincial governments to change this situation and to restore valuable forest ecosystem.

Central questions in this chapter are: What is the environmental effect of these payment schemes in Fujian Province? What are the consequences of these schemes for local livelihoods? And how do the actors participate in these schemes? Hence, in this chapter we analyze the emergence of local payment schemes and examine their performance with respect to environmental effectiveness, livelihood consequences and local participation.

This chapter starts with an overview of the institutional setting for payment schemes in Fujian province (Section 4.2). Subsequently in Section 4.3, Fujian's existing payment schemes are introduced. In Section 4.4, we report on a detailed evaluation of these payment schemes in Yongtai County, Fujian Province, followed by a conclusion in Section 3.5.

## 4.2 Institutional setting for payment schemes in Fujian

Payment schemes in Fujian are supported by classification-based forest management and influenced by the collective forest tenure reform. This section first introduces the implementation of classification-based forest management in Fujian. Then the process of collective forest tenure reform on public benefit forest is presented and its impacts on payment schemes are analyzed in detail. In addition, political willingness of Fujian provincial government is an important factor to give birth to existing payment schemes and to orient the formulation of the schemes.

### 4.2.1 Classification-based forest management in Fujian

Since 1995, China has adopted a strategy to manage forests based on two primary functions: commercial forest and public ecological benefit forest[14]. A series of regulations were subsequently issued by the central government to direct the management of public benefit forest. These regulations relate to all aspects of governing public benefit forests, including demarcation, adjustment, requisition, management, protection, logging, payment, monitoring and inspection.

---

[14] Following the *General Outline of Forestry Economic Reform* and the *Notice on Piloting Forestry Classified Management Reform*, issued by the former Ministry of Forestry, some provinces have started to zone forests for different functions.

Following these national rules, Fujian Province set up its own management institutions for public benefit forests. *Fujian Province's Management Measure on Public Ecological Benefit Forest* (Fujian Provincial Government, 2005) is at the core and stipulates formal rules for classification, protection, productive management, monitoring and inspection, and punishment for noncompliance. These rules require county governments to take responsibility in protecting and managing public benefit forest within their jurisdiction. Property right of forest has been restricted such that only limited selective logging is allowed with the approval of the provincial forestry department. This management measure complemented other regulations related to funding, tenure registration, logging norms, and adjustment and requisition rules, together constituting an institutional setting for the implementation of payment schemes.

### 4.2.2 Collective forest tenure reform in Fujian

Besides these direct management institutions set up by governments, payment policies for forest ecosystem services have been influenced by decentralization in forest governance. In China, decentralization has mainly taken place in collective forests, which account for about 60% of forestland in the country. There have been two rounds of decentralization of the ownership of collective forests. The first round started at the early 1980s, in which collective forests were allocated to farmer households as *ziliushan* (family plots) and *zerenshan* (responsibility hills)[15]. However, this reform was terminated by the central government once it became clear that most farmers cut the trees and sold the wood when they obtained the use rights of forest plots, without reforestation. The second round began after the central government issued *Decisions regarding Speeding up Forestry Development* (Chinese State Council, 2003a), in order to improve forest productivity and ecological service. This decentralization employed more flexible measures to direct tenure reform (Liu, 2008). Different forest tenure and management arrangements are provided for local farmers to choose, according to local natural and socioeconomic conditions. The owners of forest are entitled to sell and lease their land-use rights, sell their standing trees or use them as investment share and collateral. In addition, a matching reform on forestry tax and fee are launched to reduce the cost of forestry production.

Although the main objective of forest tenure reform is to improve forestry production and increase farmers' income, it had a complex impact on public benefit forests. The reform widened the gap between commercial forests profits and public benefit forests profits. Hence, illegal logging on public benefit forests occurred more frequently than before and the number of criminal cases in Fujian related to public benefit forests increased from 36 in 2003 to 59 in 2005 (Lu, 2007). And efforts and interest of local farmers largely shifted from public benefit forest to commercial forest. In addition, the reform also strengthened farmers' awareness of property rights of forestland. As Fujian Province distributed only collective forests defined as commercial forest to farmer households, several villages appealed to the government to change their public benefit forest into commercial forest. In order to deal with the emerging crisis in the public benefit forest system, Fujian province launched a reform on the management of public ecological benefit forest in 2007.

---

[15] The ownership of the land of *ziliushan* and *zerenshan* belong both to the collectives of villages. The owners of *ziliushan* have almost permanent use rights to the plots, whereas the owners of *zerenshan* have to renew their contract and share the profit with the collectives, and have limited control over harvest and sale of trees.

This reform extended decentralization into the domain of public benefit forest. But the reform on the public benefit forest is fundamentally different from that on collective commercial forest (Huang, 2006). Firstly, the reform left the collective ownership of public benefit forest untouched, while in commercial forest ownership was transferred to farmer households. Secondly, whereas the commercial forest reform empowered farmer households with complete management rights, on public benefit forests only limited management rights were transferred to farmer households. These households are constrained in using forest resources without timber logging and logging for the purpose of forest tending and regeneration. Thirdly, the public benefit forest reform includes the distribution of compensation payment and responsibility of protection; obligations and benefits moved from villages to individual farmer households.

Three models have been developed by the provincial government for distributing management rights of public benefit forests, which includes the responsibility of protection and management and payment for compliance (Fujian Provincial Government, 2007). The first model distributes management rights to farmer households and organizes individual farmer households into united management teams. The payment is also distributed to farmer households, be it on an average basis. And the revenues from restricted use of a public benefit forest are evenly shared between farmer households. This model is mainly applied in regions where farmer livelihood relies strongly on the public benefit forest. The second model, applied in regions where farmers are less dependent on public benefit forests, contracts out the responsibility of forest protection to local farmers or a third party following an open procedure. The payment by governments is distributed equally among farmer households, after paying the rangers who take care of forest protection according to contract. The distribution of revenues from restricted use is determined by a village conference or a representative conference. Not unlike the second model, the third one allows state-owned forest farms to take care of all collective forests for public benefit. This model is applied when public benefit forests are small in size and difficult to manage and protect.

### 4.2.3 Political willingness of government for PES

Fujian has a strong motivation and political determination to maintain and improve its environmental and ecological quality. In his report on the government's work 2008, the governor, Mr. Huang Xiaojing, urged his administration to continuously focus on energy saving, pollution control and eco-compensation in watersheds (Huang, 2009). There are important institutional factors which facilitate agenda setting and formulation of payment schemes (especially those at provincial level) in Fujian Province. Fujian's forestry administration plays an important role in initiating local payment schemes, negotiating with other governmental agencies, and pushing for payment policymaking. In Fujian, the provincial forestry department has a relatively high administrative status in the government, compared to its counterparts of other provinces such as Guangxi and Qinghai. Fujian provincial forestry department as one component of the provincial government has more access to policy decision of the government than provincial forestry bureaus in other provinces. It can also relatively easily influence political agenda setting. In addition, forestry is an important sector in Fujian's rural economy. The livelihood of local farmers is still largely dependent on forestry income. This importance of forestry also increases its influence on political agenda setting and capacity to mobilize public financial resource for forest management.

The policy process of payments from downstream to upstream areas exemplifies how the forestry department uses power and resources to mediate between demanders (downstream regions) and suppliers (upstream regions) for forest ecological services within its jurisdiction. In spite of the strong capacity of the forestry department, the process was not smooth but was accompanied with compromises and negotiations. At the beginning, the forestry department suggested a policy arrangement that would directly collect funding from industries which benefit from forest ecological services, such as dams and hydropower stations. However, water conservation and hydropower management departments strongly opposed this arrangement, because they have close relations with these industrial sectors. Facing this setback, the forestry department changed its strategy to request funding from public fiscal budget of the provincial government and downstream cities rather than from specific industries. The proposal was accepted by all provincial departments but opposed by downstream municipal governments. Then the provincial forestry department reformulated the proposal which would pool funding from all cities according to water consumption amount but distribute the payment by area of public benefit forests. As a result, upstream cities obtain a larger share of funding than downstream cities. Finally, this proposal was approved by the provincial and municipal governments and the payment scheme was established.

Another governmental agency – the Provincial Environmental Protection Bureau (PEPB) – is also involved in the policy domain of PES. Together with Fujian Financial Department, Fujian Development and Reform Commission and their local counterparts, Fujian Provincial Environmental Protection Bureau (PEPB) initiated, financed and implemented an eco-compensation scheme in Min River watershed. It started from 2005 and anticipated to integrate all PESs for watershed including forest related PESs in future. To this end, on behalf of the provincial government, Fujian PEPB drafted the *Proposal on Strengthening Water Environment Protection in Min River Watershed*, which was circulated among departments at provincial, municipal and county levels in March 2009 and consultation meetings were organized with these departments. It helped to raise awareness of eco-compensation among different governmental agencies and promised to improve the sustainability and the legitimacy of the schemes in future.

## 4.3 Payment schemes in Fujian Province

Payment for forest ecosystem services in Fujian province comes from four sources: the national payment scheme, the provincial payment scheme, payments from downstream to upstream areas, and local payment schemes. The four schemes provide payments for the owners of all so-called Public Benefit Forest in the province. I will give a brief introduction on these schemes below.

The Forest Ecological Benefit Compensation Fund program (FEBCFP) is the only national payment policy carried out in Fujian Province. Before 2001, *Regulation on the Implementation of the Forest Law* (Chinese State Council, 2001) stipulated already that

> *the area of a key shelter forest or special-purpose forest [these two types of forests have been classified into public benefit forest] within the administrative area of a province, autonomous region, or municipality directly under the Central Government shall not be less than 30% of the total forest area of the said administrative area.*

But the central government did not provide special funding for public benefit forest management, let alone compensate economic loss of forest owners following logging restrictions. To encourage forest owners to protect and restore forest ecosystems, the central government experimented with FEBCFP in 2001 and extended the coverage of the program and increased the payment standard gradually. As one of the pilot provinces, Fujian province delineated 2.86 million ha as public benefit forest in 2001 (30.7% of its forestland). And the central government and provincial government shared the responsibility to provide financial support for the public benefit forest. From the start the central government supplied 65 million Yuan for 8,666.67 km² of public benefit forest annually (see Table 4.1). Generally, for centrally financed public benefit forest, the payment standard is 7,500 Yuan per km²: 6,750 Yuan for forest owners and 750 Yuan for public expenditures such as fire and pest control.

Parallel to the central payment which only covered 30.27% of the public benefit forest, Fujian provincial government provided initially 57.20 million Yuan annually for the payment for the rest of the public benefit forest. Although the standard of the provincial fiscal payment for public benefit forest was only 2,025 Yuan per km² in 2001, it rose to 7,125 Yuan per km² in 2007, close to the central payment level. According to the "Measure on the Management of the Central Forest Ecological Benefit Compensation Fund" (Ministry of Finance and State Forestry Administration,

Table 4.1. Funding sources for forest payment schemes in Fujian Province, 2001-2007.[a]

| Year | Central government | | Provincial government | | Downstream to upstream (million Yuan) |
|---|---|---|---|---|---|
| | area (km²) | payment (million Yuan) | area (km²) | payment (million Yuan) | |
| 2001 | 8,666.67 | 65.0 | 19,933.33 | 57.2[b] | - |
| 2002 | 8,666.67 | 65.0 | 18,600.00 | 30.0 | - |
| 2003 | 8,666.67 | 65.0 | 16,666.67 | 34.0 | - |
| 2004 | 8,666.67 | 65.0 | 16,466.67 | 50.1 | - |
| 2005 | 8,666.67 | 64.5 | 16,600.00 | 65.2 | - |
| 2006 | 13,800.00 | 103.4 | 14,866.67 | 110.2 | - |
| 2007 | 13,800.00 | 103.4 | 14,866.67 | 121.3 | 85.9 |

[a] Provided by the Planning and Financial Division, Fujian Provincial Forestry Department.

[b] It is quite general for national forestry projects to set a local matching requirement for provinces in China. The purpose is to strengthen the responsibility of local governments to implement the projects and control their desire to include too many forestlands into the projects in order to acquire more funds from the central government. Therefore, in 2001, Fujian Province contributed a high provincial fund to match the national fund, in order to compete to apply for larger public benefit forest in its own jurisdiction. At that time, the central government also showed the will to raise the standard for compensation, but the promise has not been realized until 2009. After 2001, with the transfer of policy focus, the provincial government reduced its contribution for the local fund.

2004), local forestry departments are required to make a contract with state-owned forest farms and villages, concerning forest management responsibility and payment distribution. Also villages have to enter into a contract with forest rangers in case the forest is collectively owned or with farmers who own the forest individually. Moreover the province promoted two rules for public benefit forest management: the owners of the public benefit forest have the rights to obtain payment for the ecological service they supply; and local governments should share the payment responsibility with the central government.

In 2007, a new payment scheme "Forest Ecological Compensation from downstream regions to upstream regions" emerged to create a new financial source for supporting the public benefit forest. Under this payment scheme, nine cities contributed to a common compensation fund, mainly based on the amounts of their urban and industrial water in 2005. The contribution will be measured each three years. The provincial government collects all contributions from municipal governments and redistributes the funds to them according to the area of public benefit forest within their jurisdictions. In 2007, this scheme collected 85.90 million Yuan and provided a payment of 3,000 Yuan per $km^2$ for all public benefit forests within the province (see Table 4.1).

Besides these central and provincial initiated payment schemes, municipal and county governments also began to facilitate diverse payment schemes for forest ecosystem services at a small scale. These payment schemes fall into three types: public fiscal support, payment from direct beneficiaries, and ecological fees on timber production. Some counties in Fujian province have arranged funds from their county fiscal budget to complement the central and provincial payment. For example, Shaxian County's finance department provides 200,000 Yuan for the payment for public benefit forests from its own fiscal budget. Apart from public fiscal support, the county governments also raise money from companies and individuals who directly benefit from ecological services provided by public benefit forests. In Yongtai County (our case study area), 1% revenue of hydroelectric power stations and 3% of forest related tourism revenue have to be contributed for the payment fund. In addition, in counties with a large wood processing industry part of the timber production fees flows into payment fund.

In addition, the eco-compensation scheme for Min River watershed was initiated by Fujian Provincial Environmental Protection Bureau (PEPB). This scheme sets targets for water quality from 2009 at river sections along the main stream and the tributaries of Min River and the functional zones. For instance, over 95% of the sections monitored by Fujian PEPB should meet class III, and over 95% of the drinking water sources for cities should meet the quality standard. To achieve these targets, tasks ranging from drinking water sources protection, animal waste treatment, household waste treatment, industrial pollution control to watershed ecology protection are listed and assigned to relevant departments. It calls for the establishment of Leading Group for Water Environment Protection in Watersheds by the provincial government to be responsible for overall coordination and supervision. Fujian PEPB and Fujian Financial Department are called to formulate concrete measures for PES schemes. This scheme just started in 2009 and has not yet included forest-related components. Our investigation in this province does not include this scheme.

The diversification in funding payment schemes from 2007 onwards came with controversies. For instance, the leader of the General Management Station for Forest Resources in the Fujian Provincial Forestry Department (PFD) stated that beneficiaries of forest ecological services, such as downstream regions, should contribute much more to the payment schemes. Local officials from the Sanmin Municipal Forestry Bureau (MFB) saw central and provincial governments as mainly responsible for funding the schemes and they expressed their worry that, if municipal and county governments developed local payment schemes, the central government would severely reduce funding support.

## 4.4 Evaluating payment schemes for environmental services

In the assessment of payment for forest services in Fujian province the framework of Chapter 2 is used, where the institutional setting influences payment schemes and payment schemes influence forest use practices of local farmer household, potentially resulting in changes in forest ecological services, farmer livelihoods and participation. A before-after case study was used to understand implementation of payment schemes and to examine their impacts on forest ecological services and local farmer livelihoods. One county in Fujian, Yongtai, has been selected as case study area. We will first introduce the operationalization of the three main variables (forest ecological services (environmental effectiveness), livelihood and participation), then introduce the case study and finally outline the data collection methods.

Measuring environmental performance – the most important objective for payment programs – is difficult, especially for forest protection programs, for 3 reasons. Firstly, there exist a major time lag between protection activities and environmental outcomes. Secondly, besides payment schemes many intervening factors may influence the outcome, which cannot be ruled out easily. Finally, it is difficult to acquire adequate data on environmental quality, especially in developing countries. In this study a first rough assessment is made on the environmental effect of PES, using two data sources for ecological quality changes of forests: observations of local farmers measured through household questionnaires, and forestry resource inventories on the quality of the forest ecosystem, using 5 indicators (see below). The impact of payment schemes on the livelihood of farmer households is analyzed in this study through household questionnaires and interviews with officials, focusing on income changes of farmer households; the impact on firewood collection, livestock agriculture and cash cropping in forests; and effects on other income sources for communities (especially tourism). Participation focuses on the assessment to what extent and how targeted recipients have participated in the formulation and implementation of the payment scheme. Household questionnaires, interviews and group meetings have been used to collect data both on the actual participation at the different stages of the payment programs and on the attitude and comments of farmers and local officials on the participation mechanisms of payment programs.

### 4.4.1 Introduction to the case site

To understand and assess the implementation of payment schemes a typical case study area in Fujian province has been selected, taking into account the following criteria: the starting time of payment schemes (the earlier, the better), a large ratio of public benefit forest, forests with high ecological and economic importance, existence of forest tenure reform, and reforms in public benefit forest management. Yongtai County proved to be an excellent case study area, fulfilling best these criteria. Yongtai County lies southwest of Fuzhou City, the capital of Fujian Province. Its jurisdiction includes 21 townships, 2,241 sq. km and a population of 360 thousand. Yongtai County is located in mountainous region with forest coverage of 76.8% (Figure 4.2). The public benefit forest (40.0% of the forestland in county) provides two major ecological services: watershed protection for drinking water for Fuzhou and other major cities; and landscape value for several forest tourism sites. The county still has a sizeable forest industry including timber harvesting, timber and bamboo processing, and bamboo and rattan weaving. In 2007, the total production value of the forest industry reached 981 million RMB (22.76% of the county's GDP).

Five villages were selected from two townships using similar sampling criteria as for the county (Table 4.2). All five villages are located in mountainous areas but have good access (with cement

Figure 4.2. The mountains in Yongtai County, Fujian Province.

Table 4.2. Socio-economic situation of sample villages in Yongtai County.[1]

| Township | Village | Land area (ha) | Households | Population | Public benefit forest area (ha) |
|---|---|---|---|---|---|
| Linglu | Changkeng | 2,204 | 92 | 430 | 1,766 |
| Linglu | Yunshan | 1,151 | 135 | 608 | 909 |
| Linglu | Zhaixia | 1,246 | 100 | 452 | 1,044 |
| Baiyun | Baiyun | 858 | 436 | 2,057 | 363 |
| Baiyun | Zhaili | 1,019 | 365 | 1,365 | 210 |

[1] Source: Yongtai County Forestry Bureau.

roads) to towns and cities. Changkeng Village has the largest area and the smallest population size, while Baiyun Village has the largest population size and the smallest area. The other villages (Yunshan, Zhaixia and Zhaili) have almost the same area but differ in population densities. Two villages (Zhaixia and Zhaili Village) have frequent occurrence of rainstorms (>5 times per year); the occurrence of rainstorm in the other three villages is normal (2-5 times per year). Changkeng has the highest income per capita and Zhaixia is the poorest village (Table 4.3).

### 4.4.2 Data collection

*In-depth interviews, group meetings, survey*

In-depth interviews, group meetings and surveys have been employed to collect information and data. In understanding the payments schemes and developing survey questionnaires, in-depth open interviews have been conducted with officials in the national and provincial forestry departments, officials at the county and villages (especially in Yongtai County, see below), and with local farmers and forest owners. Two group meetings were organized, one for local forestry officials and the other with heads of 5 villages and local farmers (Figure 4.3).

After the interviews, a survey of farmer households in 5 villages of Yongtai County was conducted during August 2008 (Table 4.4). Between 9 and 11 farmer households have been randomly sampled in each village. All respondents, the heads of the households, were male and Han people. A total of 50 farmer households were interviewed and 49 useful surveys were obtained. The farmer household survey asked farmers about their income and livelihood, participation in payment schemes, attitudes toward payment regulations, and assessment on the performance. Four respondents (out of the 49) refused to reveal their household income. The lack of young respondents in the survey is attributed to the fact that young family members usually take off-farm work outside the villages or even the province. All respondents completed primary education and 63% finished middle school education. 76% of interviewed families have 4-6 members and 57% of the families have 2 children. The general family pattern in the villages shows that one or two young members have off-farm work and old family members take care of children and engage in agricultural production.

*Table 4.3. Income distribution in villages in Yongtai county (in RMB).[1]*

| Township | Village | <637 | 637-900 | 901-1,500 | 1,501-2,500 | 2,501-3,500 | >3,500 |
|----------|---------|------|---------|-----------|-------------|-------------|--------|
| Linglu | Changkeng | 0.0% | 0.0% | 2.0% | 8.0% | 20.0% | 70.0% |
| Linglu | Yunshan | 0.5% | 1.5% | 3.0% | 10.0% | 70.0% | 15.0% |
| Linglu | Zhaixia | 3.0% | 5.0% | 10.0% | 60.0% | 15.0% | 7.0% |
| Baiyun | Baiyun | 0.0% | 0.0% | 0.0% | 30.0% | 70.0% | 0.0% |
| Baiyun | Zhaili | 0.0% | 8.0% | 12.0% | 13.0% | 62.0% | 5.0% |

[1] Source: questionnaire at village level 2008.

*Figure 4.3. A group meeting with local farmers at Linglu Township.*

*Table 4.4. Characteristics of farmer households Yongtai county.[1]*

| Residence | Changkeng | Zhaixia | Yunshan | Baiyun | Zhaili |
|---|---|---|---|---|---|
| | 11 | 9 | 10 | 9 | 10 |
| **Age** | **30-39** | **40-49** | **50-59** | **≥60** | |
| | 11 | 21 | 12 | 5 | |
| **Education** | **Senior** | **Junior** | **Primary** | | |
| | 9 | 31 | 9 | | |
| **Family size** | **≤3** | **4-6** | **≥7** | | |
| | 6 | 38 | 5 | | |
| **Children** | **≤1** | **2** | **≤3** | | |
| | 13 | 28 | 8 | | |

[1] Source: farmer household survey 2008

In addition to in-depth interviews, group discussions and a farmer household survey, the heads of the five villages were asked to complete questionnaires for their village, with questions on their personal information, environmental and socioeconomic background of the villages, village involvement in payment schemes, forest tenure reform, effects of the schemes, etc. All five village level questionnaires have been finished completely.

*Forest resource inventories*

In addition, forest resources inventories – both before and after the introduction of payments schemes – on 10 farmer household plots of the public benefit forest in Yunshan Village were collected from the local forest stations in order to evaluate whether payment schemes have improved the quality of ecological services. Indicators (Appendix E) for the forest examination come from the national standard Non-commercial Forest Construction Guide Principle (GB/T18337.1–2001), which has been widely used to measure the quality of public benefit forests.

### 4.4.3 Performance of the payment schemes of Fujian province

The evaluation for payment schemes includes three parts: environmental effectiveness, livelihood impact assessment and participation mechanism analysis.

*Environmental effectiveness*

According to the *Planning Outline of Fujian Provincial Forest for Ecological Public Benefit* (Fujian Provincial Forestry Department, 2005), the forests in Yongtai County are included in the Minzhong and Minnan (South and middle of Fujian) Public Ecological Benefit Forest Zone, where typhoons and soil erosion are the main ecological threats. However, the environmental degradation is not evenly distributed within the zone, due to various geographic factors and human intervention.

According to the farmer household survey, 55% of the farmers believed their villages had no apparent soil erosion before the payment schemes started. About 29% of respondents thought their village had some soil erosion and only 8% thought that widespread soil erosion existed before the schemes. All respondents in Changkeng and Yunshan Village indicated no apparent soil erosion before the schemes, while the respondents of Zhaili Village all thought that the village had widespread or partial soil erosion. In Zhaixia Village and Baiyun Village opinions of the farmers were most diverse. These results corresponded with rainstorm frequency in the villages.

Around 35% of all respondents thought that the soil erosion had been significantly reduced since the introduction of the schemes; 29% of them thought that some reduction had occurred and 22% saw no change in local environment following payment schemes. With different baselines (degree of soil erosion before the schemes) assessments of ecological improvement through schemes differed. Almost all the respondents of Changkeng Village, where no apparent soil erosion had previously existed, thought that schemes brought no change to the local environment, but seven out of ten respondent farmers of Yunshan Village (no apparent soil erosion before) witnessed slight decreases in soil erosion since the payment schemes started. In contrast, all the respondents of Zhaili Village thought soil erosion has significantly been reduced, and 7 out of 9 respondents of Zhaixia Village affirmed significant or slight reduction on soil erosion after the payment schemes. Therefore, from the perspective of local farmers, the payment schemes achieved greater environmental effectiveness in the villages with severe soil erosion than those not threatened by the ecological degradation.

This environmental effectiveness outcome was compared with the forest resource inventories (Table 4.5). These inventories show that before the payment schemes the forest ecosystem in Yunshan Village was in a poor condition under which it cannot effectively control soil erosion. The average score for the 10 selected forest plots was 53.0 points (the national standard for public benefit forest controlling soil erosion is 60 points). Canopy density and vegetation coverage of the forests before the payment schemes both fell into grade III and dragged down the total score. The indicators for the post-payment period demonstrate that all qualities of the public benefit forests have been improved. The average total score of 76.6 points means that the quality of the public benefit forests changed from poor to moderate condition and is eligible for controlling soil erosion. Conditions of biodiversity and phytocenosis structure are excellent, but canopy density and vegetation coverage are still relatively low.

Hence, both methods suggest that the payment schemes increased the qualities of public benefit forests for soil erosion.

## Livelihood and compensation

Payment schemes impose limitations on the economic use of forest resources, but do provide compensation. Payments initially came from the central and provincial government, but since 2007 also from local sources (see Table 4.1). The 2007 *Management Measure for Compensation Funding for Public Benefit Forest* (Ministry of Finance, 2007) stipulated that at least 50% of the payment should be distributed to forest owners. Table 4.6 reports our findings of the distribution of payments among village committees, full-time foresters, and local farmer households in 2007. To what extent compensation can offset the loss of farmer household income is crucial for the sustainability of the payment schemes. The baseline for comparing loss and compensation was set in 2001, the year before the implementation of the first nationwide payment scheme.

According to the farmer household survey, 71% of the farmers thought that the payment schemes had not affected their income; 10% believed their income had increased with the schemes and 14% asserted that the payment schemes had a negative impact on their income. Most respondents indicating negative income effects by the schemes came from Zhaixia Village,

*Table 4.5. Average scores for the quality of the public benefit forest ecosystem.*

| Indicators | Biodiversity | Canopy density | Phytocenosis structure | Vegetation coverage | Litter layer | Total |
|---|---|---|---|---|---|---|
| Before the schemes | 11.5 | 8.3 | 13.0 | 8.5 | 11.7 | 53.0 |
| After the schemes | 19.0 | 11.2 | 19.0 | 11.6 | 15.8 | 76.6 |

*Table 4.6. Distribution of payments within villages, 2007 (%).[1]*

| Villages | Village committee | Foresters | Farmer households |
|---|---|---|---|
| Zhaili | 15 | 20 | 65 |
| Changkeng | 15 | 30 | 55 |
| Baiyun | 15 | 35 | 50 |
| Yunshan | 15 | 30 | 55 |
| Zhaixia | 15 | 30 | 55 |

[1] Source: interviews village level questionnaire.

accounting for 67% of the village respondents. Especially in Yunshan, Baiyun and Zhaili almost all respondents found their income had not been affected (Table 4.7).

According to the survey, the average household income decreased slightly by 7.0% from 12,133 Yuan in 2001 to 11,282 Yuan[16] in 2007. On average, off-farm work composed the main source of household income both before (6,313 Yuan) and after (5,752 Yuan; 51% of total income) the payment schemes started. However, average household income from agricultural production decreased substantially from 2,974 to 1,689 Yuan between 2001 and 2007. Although the payment schemes might provide incentives to transfer rural labor from forestry production to off-farm work, the survey did not show a strong direct relation between the payment schemes and rural labor transfer.

*Table 4.7. Farmer assessment on income impacts of payment schemes.[1]*

| Village | Increase | Decrease | No change | Don't know |
|---|---|---|---|---|
| Changkeng | 4 | | 7 | |
| Yunshan | | | 10 | |
| Baiyun | | 1 | 7 | 1 |
| Zhaili | 1 | | 8 | 1 |
| Zhaixia | | 6 | 3 | |
| Total | 5 | 7 | 35 | 2 |

[1] Source: farmer household questionnaire 2008.

---

[16] The monetary value of farmer household income in 2007 have been adjusted into real terms (taking 2001 as the base year), according to the consumption price index for rural residents provided by China Rural Statistics Yearbook 2010 (National Bureau of Statistics, 2010).

Before the payment schemes, the average household income from timber logging was 148 Yuan in 2001 and accounted only for 1.2% of the total household income. After introduction of the payment schemes timber harvesting disappeared almost completely and income decreased to 7.5 Yuan in 2007. While restrictions of payment schemes on timber harvesting had no significant impact on average household income of farmers, the survey showed that 4 farmer households depended for over 15% of their income on timber harvesting in 2001 (and one even for 25%). For farmers, whose forests had grown into final cutting age[17], restrictions imposed by the schemes can thus have a considerable negative income impact. From the village-level questionnaires, this problem proved prevalent in Zhaixia Village, where 68.9% of forests had been used for timber production and already entered into final cutting age and we also see most respondents indicates decrease of income. Therefore, on a long term of 10-20 years, the income loss of local farmers from the compensation schemes can be relatively large.

Apart from the direct income loss, the forest use restrictions of the payment schemes also had negative impacts on daily fuel and other agricultural production, such as reduced supply of firewood and prohibition of livestock browsing on forestland (Figure 4.4). Over 32% of the respondents thought that the schemes reduced fuel wood consumption, 14% indicated that their family animal husbandry was affected and 20% mentioned other difficulties in their daily life due to the schemes.

But the restricted use of the public benefit forest improved the development of local tourism. During the group meeting, the local officials stated that the county had developed a sizable tourism due to a good forest ecosystem. Fifteen eco-tourism areas have been established within the public benefit forests, which covered 10.5% of the total area of public benefit forest in the county. In 2007, 983,000 tourists visited the county, and the total market value of tourism was 278 million Yuan.

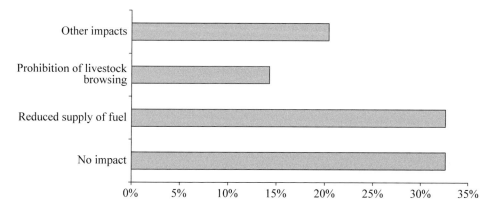

*Figure 4.4. Payment schemes' impact on non-forestry activities of local farmer (n=49).*

---

[17] For timber production, the final cutting age for the broadleaved trees in Fujian Province is 20 years, which means at this age, trees can be cut down and produce the most amount of timber. So farmers have more desire to cut down trees for timber production at the final cutting age.

Village committees that opened the public benefit forest for ecotourism received forest resource use charges from local tourism companies[18]. Zhaixia village, for instance, obtained annually 9,000 Yuan from tourism companies for using the public benefit forest at Shuiliangong Scenery Spot. These forest use revenues from tourism companies are usually managed by the village committee and used for public services such as road maintenance, education and public facilities. These charges do not reach farmer households in cash. Interviewees indicated that some farmers benefited from tourism development by running small family hotels and restaurants, working for tourism companies, and/or selling local specialties. However, data from the farmer household survey did not prove this, perhaps due to small sample sizes and the specific locations of these profits.

Farmer households received payments from the government, a direct benefit for them to participate in the payment schemes. Before 2006 farmer households had not received any payment from local governments. In 2006, 16.3% of farmer households received payment from the schemes, between 181 and 545 Yuan per household. The coverage of the payment increased dramatically to 79.6% in 2007, with payments up to 1,600 Yuan per household. Four villages – Baiyun, Yunshan, Zhaixia and Changkeng – equally distributed payment per capita (including kids). At first, the village committees divided the payment into three parts. The first part is the payment for every village member as compensation, since they are the collective owners of the public benefit forest. The second part is the payment for some members who take the responsibility of managing and protecting the forest. The third part is for the village committees for pest and fire control. The level of payment depends on the area of public benefit forest and the number of village members. Baiyun Village had the lowest level of payment with 4.3 Yuan per capita per year. The payments in Yunshan and Zhaixia Village were at modest levels with 69 Yuan per capita in Yuanshan and 129 Yuan per capita per year in Zhaixia. Changkeng Village had the highest payments with about 224 Yuan per capita per year. Different from other villages, Zhaili Village only distributed payments to farmer households which lived closed to the forests. The households of Zhaili received payment according to the area of the forest under their management/tenure. The standard for the payment was 6,033 Yuan per km² annually.

There are two reasons for the increase of the amount and coverage of the payment. Firstly, the forest tenure reform and its matching reform – the management reform on the public benefit forest –strengthened the perception on the property rights of forest and entitled each village member to benefit from the collectively owned forest. This change facilitated the redistributing of payments to local farmers. Secondly, in 2007, the provincial payment scheme "Forest Ecological Compensation from downstream regions to upstream regions" was established to channel extra funding to the public benefit forest. This enabled local governments to redistribute the payment to farmer households, as before 2007 the meager payment from the central government was often captured exclusively by county forestry departments, township governments and villages.

On average, the payment only accounted for 3.2% of the household income of farmers in 2007, ranging between 0% and 16.67% for the sampled households. The proportional and absolute payments had an equal uneven distribution among the villages. The high payment in Changkeng

[18] According to the group meeting, Xiyang Village, which is located in the Tianmen Mountain Scenery Spot, has been annually getting 6,000 Yuan from the local tourism company – Tianmen Mountain Scenery Management Company for using the public benefit forest.

Village matched with the fact that about 40% of the respondents thought the payment schemes increased their income. However, farmers in Zhaixia had also a proportionally high payment (about 6%), but largely thought they had not been fully compensated by the payment schemes.

### Participation in formulation and implementation

This part discusses how farmers participated in policy formulation and implementation of payment schemes, concentrating on four stages: policy formulation, demarcation, management and examination of payment schemes.

a.  Payment policy formulation. In developing payment schemes, the goal of the provincial and the central government was demarcating sufficient forests into public benefit forest. Local farmers and village committees were mainly focused on reducing economic losses. The village committee and local farmers used different channels to express their opinion and to influence county forestry bureaus and local officials. County forestry bureaus often forwarded the general opinion of local farmers on the payment schemes to the provincial forestry department. Local officials, when elected as delegates of Provincial People's Congress (PPC)[19] or selected as members of Provincial People's Political Consultative Conference (PPPCC), can propose suggestions during the Two Meeting each year. For 2004 to 2006 major proposals were made, related to payment standards, adjustment of public benefit forest areas, local payment schemes, and rights of forest owners (see Table 4.8). The policy recommendations in 2005 and 2006 directly gave birth to the Reform on the Management of Public Benefit Forest in 2006 and the Payment Scheme "Forest Ecological Compensation from downstream regions to upstream regions" in 2007. However, not all proposals are accepted and transformed into new institutions and policies. For instance, the provincial forestry department refused the suggestion to re-plan and re-demarcate the area of public benefit forests, as suggested in several proposals. The Forest Law stipulates a strict proportion of public benefit forests (at least 30% of total forest) for each province, giving the provincial forestry department little room to reduce the total area of public benefit forests. It did allow tiny adjustments of public benefit forests under special situations[20].

b.  Demarcation of public benefit forest. Payment schemes for forest ecosystems generally involve four parties: the central government, local governments (provincial government, county governments and township governments), village committees and local farmers. The *Measures on Demarcation of Key Public Benefit Forest* (State Forestry Administration and Ministry of Finance, 2004) stipulates the principles for demarcating public benefit forests: "the demarcation process should respect the rights of forest owners and managers, keep forest tenure stable and

---

[19] PPC is a forum for mediating policy differences between different parts of the government and the regions, and PPPCC is a political advisory body including broader members such as democratic parties, persons without party affiliation and mass organizations. Both are important institutions for public policy making and consultation.

[20] A prior consideration of adjustment can be given for private forest owned by individual farmer households (Ziliushan), if the farmer households strongly asked to quit the payment schemes; if 80% of collective forest land in a village had been demarcated into public benefit forest, the proportion can be reduced to 50%; if owners of plantation had refused signing demarcation agreement or accepting the payment, their plantation can be considered for adjustment.

*Table 4.8. Number of proposals relating to payment policies during PPC and PPPCC meetings of Fujian Province, 2004-2006.[1]*

| Year | PPC | PPPCC | Major issues |
|---|---|---|---|
| 2004 | - | 4 | increase payment standard |
| 2005 | 15 | 9 | increase payment standard |
|  |  |  | adjust PBF areas |
|  |  |  | establish local payment schemes |
|  |  |  | protect rights of forest owners |
| 2006 | 10 | 4 | increase payment standard |
|  |  |  | adjust PBF areas |
|  |  |  | compensate villages for economic loss |

[1] Source: archives Forestry Department Fujian.

maintain existing contract of forest; local governments should make contracts for limitations on logging with forest owners on a voluntary basis; if forest owners oppose to demarcate their forest into public benefit forest, local governments should deal with the problem in an appropriate way". Hence, there is a consultation requirement in the process of demarcation, where forest owners should be informed on the demarcation and can argue against the decision; but local governments take the final decision.

In implementing these central policies, provincial governments – including Fujian Province – detailed measures for the demarcation process, but neglected principles of consultation. As a result, Fujian had no specific participation requirement for the 2001 demarcation process. Officials in the provincial forestry department explained that the tight time schedule resulted in reluctance of governments to allow participation of local forest owners. Officials of county forestry bureaus claimed that participation of local farmer households and villages is not important in the demarcation process, as it should solely rely on technical standards. Sufficient compensation for forest owners was believed to be the only criteria for accepting demarcation decisions by forest owners.

Four of the five village heads of our sample thought that the area of public benefit forest should be decided though a village meeting involving all members; one village head thought that the village committee should make the decision. In Fujian Province, village members elect village committees but the members of village committees also receive an allowance from township governments (in Yongtai County, a head of villages receives 300 Yuan monthly). On the one hand, village committees represent the interest of all members and manage collective properties including all collective forests; on the other hand, the committees need to cooperate with the township government and accept tasks such as implementing payment schemes, distributing payments and negotiating with individual forest owners. Therefore, the committees bridge local governments and farmers and constantly balance interests of their members and of the government. This structure renders the state legitimacy and governability in China's rural areas (Wang, 1997), which is crucial for smooth implementation of public policies, including payment schemes. The farmer household survey showed that 33% of the farmers thought that the demarcation should be decided by the village committees and 35% by the village meeting. Preferring village committees above formal village meetings illustrate the trust of local farmers in village committees, since a village meeting is generally required for important decisions on main interests of village members.

In 2001 the provincial department exclusively demarcated public benefit forest by using technical standards. After demarcation, the provincial government informed county governments and then township governments. The township governments subsequently negotiated the demarcation arrangement with village committees. If opposition was absent in committees, contracts on public benefit forest management were made between township governments and village committees, usually after village committees informed/consulted their members. Village committees informed or consulted 94% of the farmers interviewed on demarcation. All villages had held a representative meeting for consultation and 86% of the respondents had participated in such meetings. Following our survey, 46% of respondents strongly agreed with the demarcation arrangement, 42% basically agreed to it, and 12% disagreed. For the interest of ecological restoration, 96% of the farmers considered the demarcation arrangement reasonable. Disagreement mainly focused on the negative impact of the payment schemes on their income. No formal procedure existed to deal with conflicts in contract making. Some villages (outside these sampling villages), which found that the government demarcated too much of their forests into public benefit forests without sufficient compensation, succeeded in boycotting the zoning decision of the provincial government on their forests.

While local farmers were unable to participate in designing, modifying or fine tuning the demarcation arrangement, processes of informing and consulting local farmers rendered the schemes legitimate and acceptable for most farmers. Negative impacts of schemes on local livelihoods cannot always be offset easily by higher compensation, because local lifestyle of using timber as construction materials and daily fuel cannot be changed immediately, and too high compensation standard decreases scheme efficiency.

c. Management of public benefit forest. Management of public benefit forest includes signing management contracts, distributing forest protection tasks and conducting daily forest management. There are three types of contracts for village committees. The first one is the administrative responsibility contract between township governments and village committees. Once the forest is demarcated as a public benefit forest, the township government will make a administrative responsibility contract with village committees to transfer the responsibility of the public benefit forest management to village committees. The survey showed that 81.0% of farmers had been informed in the signing of responsibility contracts.

The second type is the management contract between village committees and full time foresters, which could be local farmers or external organizations and individuals. Through the management contracts, village committees pay foresters annual wages for forest protection. All 5 sample villages organized a meeting of village representatives to decide on the employment of foresters, and 96% of farmers were consulted before deciding on employing foresters. Farmers indicated that they completely (54%) or basically (46%) agreed on the choice of foresters by the village committee.

The 2006 reform on the management of public benefit forests resulted in a third contract type between village committees and individual farmer households. This contract established a mechanism to distribute compensation payment to farmer households and shared the responsibility of forest protection among farmers. The survey showed that 90% of the farmers signed such a contract with the village committees. Most farmers (84%) found the area of the public benefit forest under their management reasonable, while 16.3% thought they got a too limited area (and thus too little payment). Furthermore, only 21% of the farmers considered payment appropriate, whereas 79% judged the payment as too low.

d.  Examination on the performance of the payment schemes. According to Fujian Province *Management Measures on Public Benefit Forest* (Fujian Provincial Government, 2005), county forestry departments have the responsibility of organizing township forestry stations to examine performance of protection and management of public benefit forests. The results of self-examination are reported to municipal governments and at least 5-10% of public benefit forests are reexamined by municipal governments afterwards. The final result is reported to the provincial forestry department. The document does not mention formal requirements of public participation for performance examination.

Even without a formal participation procedure, informal consultation with farmers widely accompanied the examinations. Our survey showed that 94% of the farmers had been consulted for opinions and suggestions when forestry stations and high level governments conducted the (re)examination of public benefit forest management. Among the farmers who gave opinions on or revealed problems with payment scheme implementation, 78% directed their input to township forestry stations, 19% to county forestry bureaus and 14% to village committees[21]. Efficiency of performance examination was considered highest with forestry stations (47%), and less with combining forestry departments at different levels (25%), with county forestry bureaus (20%), or through self-monitoring of local farmers (8%).

Township forestry stations absorbed grassroots opinions on the performance of the payment schemes, and were recognized by farmers as appropriate and important organizations for monitoring and examination of the payment schemes (Figure 4.5). Though forestry bureaus and village committees are also important these are either too far from local contexts (county forestry bureaus) or too closely involved (village committees).

## 4.5 Conclusion

Fujian Province is a relatively isolated geographical area in China. Within its jurisdiction, the forest ecosystem along the upstream of rivers constantly provides ecological service to the downstream regions. Geographical closeness between upstream and downstream regions facilitates the emergence of a range of local payment schemes. Three institutional factors have prompted the development of the payment schemes in Fujian. First, classification-based forest management in Fujian offers an institutional setting for payment schemes by stipulating formal rules for classification, protection, and management of public benefit forest. Second, collective forest tenure

---

[21] Some farmers gave their advice and revealed problems to multiple organizations at different levels, including county forestry bureaus, township forestry stations and village committees.

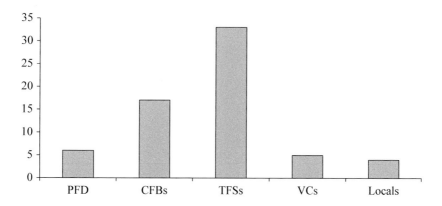

*Figure 4.5. Frequency of preference to public benefit forest examination body (n=47).*
*PFD: provincial forestry department; CFBs: county forestry bureaus; TFSs: township forestry stations;*
*VCs: village committees; Locals: local farmers.*

reform, which reshaped the structure of forest ownership in rural area, widened the gap between commercial forests profits and public benefit forests profits and thereby led to the creation of new local payment schemes to fill the gap. Finally, the political willingness of government for PES played an important role in influencing political agenda on developing local payment schemes.

The payment schemes for forest ecological services in Fujian have been designed centrally and implemented top-down. The provincial government is in charge of coordinating, funding and implementing in the Forest Ecological Benefit Compensation Fund Program (FEBCFP). In the local payment scheme from downstream to upstream regions, municipal governments started to play a role in funding and examination. However, the governments are the main implementers of the payment schemes in Fujian.

A first rough assessment was carried out within one county and the result showed that the effects of the payment schemes on environmental services is most likely positive. But payment schemes sometimes had a negative impact on the income of local farmer households, especially in regions where the public benefit forest had been used for timber production. The payments to local farmers varied in systematic and level even within one county, and did not always compensate fully the economic loss of farmers refraining from timber production.

Notwithstanding the significant role played by the central and provincial governments in the payment schemes, local forest owners/users have participated in the implementation of the schemes including their demarcation, management and examination. Village committees have played an important role in organizing local farmers to participate in the design and implementation of payment scheme and relate local farmers' needs to governmental payment scheme designs. But channels for participation of local farmers are still in need of expansion, especially in the demarcation of the public benefit forest.

# Chapter 5.
# Implementing payment schemes for forest ecological services in a poverty-stricken area – Guangxi

## 5.1 Introduction

Guangxi Zhuang Autonomous Region (Guangxi in brief) is one of the least developed provinces in China and it also has abundant forest resources. Among the southern provinces, Guangxi was the first one entering into the central payment schemes and its funding for forest protection heavily relies on the help from the central government. Therefore Guangxi is an excellent example to demonstrate the development of payment schemes in the least developed regions. This chapter first briefly introduces Guangxi's natural condition, forest resources and ecological problems relating to its forest ecosystem.

Guangxi is located in the southern part of China, consisting of 14 municipalities and 129 counties (or districts/county-level cities/autonomous counties) (Figure 5.1). Guangxi has 150,666.67 km$^2$ of forestland, accounting for 64% of the total land. The forest coverage is 54.2% according to the 2009 forest resource inventory.

Guangxi is threatened by two environmental problems, closely relating to its forest ecosystem. The first is soil erosion. The area of land under erosion[22] accounts for 11.9% of the whole region

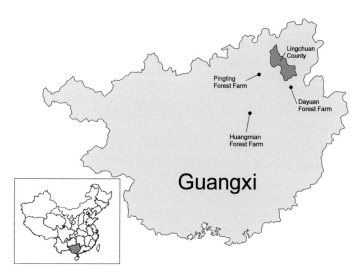

*Figure 5.1. Guangxi in China and the location of case sites.*

---

[22] The area of land under soil erosion is defined as the area where the soil erosion modulus is more than 200 tons per km$^2$ per year.

in 2009 and was more than doubled compared with the area in the 1950s and 1960s (Dai *et al.*, 2008). Construction of development projects and deforestation for cropping caused a lot of soil erosion along the Pearl River. A vast amount of soil was brought into the Pearl River and the sediment amassed at hydropower stations and reservoirs. For example, in the Xijiang River, the sediment caused the reservoirs to lose on average 1.3% of their total storage capacity each year (Dai *et al.*, 2008). Tremendous soil erosion also has a bad impact on agricultural production at the downstream region and even affects the eco-safety of the Pearl River Delta including Hong Kong and Macau. The second problem is Karst rock desertification. Guangxi is a typical Karst area made of massive soluble limestone, where the geological factors render a low soil forming capacity and poor soil quality. Once vegetation is cleared, the re-establishment will be very difficult and the mountains will become rocky without soil and vegetation (rock desertification). Karst land accounts for 37.8% of the total land of Guangxi Region and rock desertification occurs in 28.6% of the Karst land (Guangxi Public Benefit Forest Protection and Development Task Force, 2009). Many watersheds, including the Hongshui River, the Liu River, the Zuo River, the You River and the Li River, have been impacted by rock desertification. The Karst region usually has few cropland resources. Population increase drove more forest clearing for cropland and firewood. The deforestation speeds up soil erosion and in turn lowers the soil quality. This vicious cycle made that the Karst region is among the poorest and the most ecological fragile region of Guangxi.

In 2001, Guangxi started to establish its public benefit forest system and 5,1853.33 km² of forest has been demarcated as public benefit forest, accounting for 34.8% of its total forestland (Table 5.1). Especially, 54% of public benefit forest locates in the Karst region for ecological restoration.

In the next section, the institutional setting for payment schemes in Guangxi is explained in detail. Section 5.3 analyzes Guangxi's existing payment schemes including its scope and basic rules of procedure. Section 5.4 evaluates the performance of payment scheme based on a case study carried out in three sample farms and two villages. Finally, Section 5.5 formulates the major conclusions from the assessment and suggests policy recommendation.

*Table 5.1. Features of public benefit forest in Guangxi (Guangxi Public Benefit Forest Protection and Development Task Force, 2009).*

| Importance | national (prior) | regional (general) | | |
|---|---|---|---|---|
| | 74% | 26% | | |
| **Ownership** | state-owned | collective | other | |
| | 10% | 88% | 2% | |
| **Ecological function** | water source conservation | soil erosion prevention | natural reserve | other |
| | 32% | 55% | 9% | 5% |

## 5.2 Institutional setting for payment schemes in Guangxi

Guangxi's public benefit forest is managed under two systems: state-owned forest farms and villages. These two systems have different organizational structures for public benefit forest management. In order to understand forest resource management in Guangxi, we first introduce the difference between these two systems and ongoing tenure and management reforms within the systems. At the same time, the principle of classification-based forest management and related public benefit forest management policies in Guangxi build a platform for the functioning of payment schemes. A brief introduction on public benefit forest management is given and the problems in the institutional support for payment schemes are discussed. In addition, Guangxi's forest industrial policy also imposes profound influences on the payment schemes through a series of incentives and restrictions on forest use. These influences are discussed at the end of this section.

### 5.2.1 State-owned forest farms and villages for public benefit forest management

The difference between state-owned forest farms (SOFFs) and villages in forest management is apparent, and refers to forest ownership, organization, and administrative status.

State-owned forest farms are established by the state to manage, tender and protect state-owned forest. Most of the forestland under the management of SOFFs is state-owned land, but sometimes the SOFFs also operate on collective forestland based on long-term contracts. However, villages have the ownership of forestland, although this ownership is often restricted by the forest law and regulations such as land use policy and forest logging permits and quota. As forest management organizations, SOFFs operate as professional enterprises to organize forest workers to conduct forest-related production, management and protection. Villages integrate economic and political functions but their forest management is less well organized and less professional than that of SOFFs. In addition, SOFFs have a higher administrative status than villages. Although SOFFs have the characteristics of enterprises, they are also given a status in the forestry administrative system. In general, SOFFs are under the provincial or county governments and correspondingly they have the same administrative level as counties or townships. Villages have no official administrative status, but actually to some degree they function as a branch of township government.

Since the 1980s, SOFFs have experienced several reforms which were set out to solve the principal-agent problem inherent in them. Before the 1980s, the state expected that SOFFs could realize efficient timber production and sustainable forest management. However, SOFFs as an agent of the state had their own interests such as making more profit in short term. Therefore, SOFFs invested little in afforestation, increased employment without long term consideration and overharvested forest, since the governments would deal with their liabilities and meet the gap of afforestation investment in the end. In practice, it was impossible for the state to monitor timber production and forest management of SOFFs. As the information on timber production and forest management was asymmetrically distributed between the state and the SOFFs, the farms could easily divert from the objectives of the state and follow their own interests. In the 1980s, the government launched a reform to transform SOFFs into independent actors in a market-based economy. The government reduced the investment in them and gave managers more power to arrange the management of farms. However, as the government did not refer to the principal-agent problem, this reform was destined to fail. In order to make more profits and get political promotion, the managers introduced a lot of development projects on the farms after they obtained the power of management. Most projects failed in profit-making because the managers only made decisions based on "political achievement" (attracting outside investment was an important indicator for measuring performance of forest managers and local officials) but did not consider the market size of forest products and the production capacity of forestland. In 1997, China's SOFFs entered into overall industrial downturn. In Guangxi, 129 out of all 151 state-owned forest farms (most of them at county level) had their balance in deficit, together with a total annual loss of 121 million Yuan RMB and total liabilities of 1.49 billion Yuan RMB in 1998 (Liu, 1999). Since then, the central government set up a series of ecological restoration projects such as the Natural Forest Protection Program (NFPP) and the Forest Ecological Benefit Compensation Fund Program (FEBCFP). The restriction on timber harvesting has exacerbated the economic situation of forestry farms. Currently, a new round of reform divided state-owned forest farms into a commercial type and an ecological type. The commercial type forestry farms will be managed as enterprises with more flexible employment policy and have profit making as their priority. The ecological type forestry farms will be run as public facilities with governmental budget and have forest protection and restoration as their main task. In this case study, Pingling and Dayuan forest farms are in the transition from timber production enterprises into public benefit forest management agencies: their function on forest resources management and protection has been strengthened; most lumberjacks of farms have been retrained as forest rangers; professional teams for forest ranging have been set up; and forest management responsibility zones have been designated in the public benefit forest.

Collective forestland ownership in Guangxi experienced a series of fundamental changes. From 1950 to 1952, the Land Reform Campaign confiscated all forests owned by landlords and equally redistributed them to farmer households. Since 1954, Guangxi government promoted to reorganize rural communities into elementary cooperatives to support a socialist planned economy. The forest was still owned by farmer households, but the leaders of the collective made management decisions. The establishment of advanced cooperatives since 1956 brought private forestland ownership to an end. Rural households' land was merged to form collective property of advanced cooperatives. This collectivization went forward by merging advanced cooperatives into communes (known as people's communes). Correspondingly, the ownership of forests was transferred to communes. However, the large group size of communes (average 4,800 households in a commune) caused poor performance of communes in forestry production and the ownership of forests was slowly transferred down to production teams (the former elementary cooperatives and the later village subgroup) since the 1960s (Liu, 2001). In the 1980s, following the success of the household responsibility system in the agricultural sector, collective forestlands entered into its "Three Fixes" reform era: (1) stabilizing forest tenure by issuing forest tenure certificates; (2) distributing non-forest land to rural households as family plots; (3) introducing the contract responsibility system into forestry production and forest management (Miao and West, 2004). In 1985, the CCP's Central Committee and the State Council created 10 policy items to activate the rural economy, which included establishing a free market for timber and abolishing the state monopoly on purchases. Driven by such momentum some southern provinces copied the methods of the agriculture reform to distribute forests to farmer households. However, farmers' lack of confidence in tenure security and poor forest management institutions resulted in a vast amount of forest degradation (Liu, 2001). In 1987, the CCP's Central Committee and the State Council issued an order to stop distributing collective timber forest to farmer households, to re-establish the state monopoly on timber purchases in key counties for timber supply[23] and to implement strict regulatory controls on forest logging (Xu et al., 2006b). Historically, this formed ambiguous forest tenure rights in China's rural area. In practice, village committees represent village members to manage forests. However, village cadres have discretionary power on forest management and utilization. As a result, collective forestlands often become de facto owned by a cohort of village cadres (Xu and Melick, 2007). Furthermore, constrained forest logging contributed to a growing political crisis over the forest sector, which included mounting frustrations and protests among local farmers over the logging controls, reduced incentives to manage and protect forests and a growing income gap between urban and rural population (Xu et al., 2010). Since the early 2000s, some provincial-level initiatives on forest tenure reform emerged to encourage collectives to clarify forest ownership and to reallocate forest use rights to farmer households. From 2008, the central government started to promote collective forest tenure reform nationwide by supporting funding and formal guidelines. Motivated by the central government, Guangxi, with an important collective forested area, piloted the reform in two counties in 2008, extended it to 25 counties in 2009 and planned to finish the main body of the reform (delimitation, surveying, titling and registration of the new forest plots) in 2011. The regional government set an objective for each county to reallocate more than 80% of collective forestlands to individual farmer households on average (Guangxi Forest Tenure Reform Office, 2009). Different from some provinces, such as Fujian, which excluded public benefit forests from the reform, Guangxi regional government made a plan to extend the reform to public benefit forests. For the public benefit forest which has been allocated to households during the "Three Fix" period, the government is required to issue households two types of certificates (forestland use rights certificates and forest ownership and use rights certificates), make management responsibility contracts with them and distribute the whole payment for PES to them. The public benefit forest which is still collectively owned should be reallocated to households equally and the government should issue households certificates for forestland use rights, and village certificates for forest ownership and use rights. However, the public benefit forest located in nature reserves, forest parks and national scenic areas has been excluded temporally, since there is a lot of controversy about this part of the public benefit forest and its ecological importance requires more meticulous planning in the future.

However, the tenure reform is still an ongoing process. Ambiguous forest tenure and tension and conflicts between forest farms and villages also negatively affect the implementation of payment schemes. According to the interviews, the forest farms generally have forest land disputes with nearby villages. Most of the forest farms have been established in the 1950s and at that time the boundary between the farms and the land of villages were not demarcated clearly. No effort has been made to solve this problem for several decades. Therefore, with population increase and land shortage, the conflicts between forest farms and villages became more and more serious in recent years. Currently, the disputes already have had bad impacts on the payment schemes and the

---

[23] Provincial forestry departments designate key counties for timber supply according to the area of their forestry use land.

management of public benefit forests. For instance, under the opposition of villages, Pingling Forest Farm cannot receive payment from its 2,075 hectares of public benefit forest, accounting for 25.7% of its public benefit forest. This situation further worsened its financial situation and weakened effective management of the public benefit forest. Such disputes are often very complicated in terms of their history. For example, two forest stations of Dayuan Forest Farm included the land of a village in 1958 when the farm was set up and at the same time all village members have been employed by the farm as foresters as a precondition. Then, the forest farm was hit by the 1961 economic downturn, and had to run down its work force. The village members were fired and the land was returned to them. However, the location of the land has not been demarcated accurately. Moreover, the forest tenure certificates issued by the county in 1990 did not make a clear boundary between the forest farm and the village. Some forest plots are claimed by both the farm and the village. The village has appealed to the government for this dispute more than ten times, but it has not been settled. The village members have planted timber forest on some non-stocked land and scattered wood land, which they claimed and the forest farm had to give up effective management on these forestland in order to avoid triggering serious conflicts with villagers.

Besides by disputes, the forest land close to villages is also endangered by encroachment and illegal cutting from villagers, which is prevalent among most state-owned forest farms. The villages around the forest farms are usually stricken by poverty and largely depend on the forest. Wood from the forest is an important fuel for house heating in winter. Therefore the villagers around the forests often illegally harvest timber for daily fuel or to compliment financial needs such as children education. Under such circumstances, the forest farms cannot effectively monitor the forest close to villages and in some sense the forest farms undertook a policy of appeasement to the encroachment and illegal logging of the villagers. For example, the officials from Huangmian Forest Farm believed that a fire on the farm's forest in 2008 have been deliberately set by some villages because the foresters of the farm often stopped them from illegally cutting trees. The fire finally destroyed 3.33 km$^2$ of *Pinus massoniana* forest. In addition, some development projects, such as road construction and mining, also encroached some plots of public benefit forest in the case sites.

This study chooses SOFFs and villages as the objects for payment policy evaluation because they are not only the main suppliers for forest ecosystem services in Guangxi, but also their different forest ownership, organizational structure and administrative status have important impacts on the performance of PES.

### 5.2.2 Classification-based forest management and public benefit forest management policies

The demarcation of public benefit forests in Guangxi was based on 6 laws and governmental documents (Table 5.2). Most public benefit forest in Guangxi has been demarcated in 2001 and adjusted to some degree afterwards.

In practice, the counties demarcated public benefit forests and reported the result to municipal governments. Then the municipalities summed up the result of the demarcation and applied for validation from Guangxi Forest Inventory and Planning Institute. However, only about 85% of the demarcated public benefit forests satisfied the standards, since the remote sensing images used

*Table 5.2. Law and regulations for demarcation of public benefit forest.*

| Law and regulations | Issued by | Year | Main content |
|---|---|---|---|
| Forest Law and its detailed rules for implementation | the State Council | 2000 | Forests for protection and special purpose should be more than 30% of the whole forest area for each province or autonomous region |
| Guangxi Zhuang Autonomous Region Implementation Measures for the Forest Law | Guangxi People's Congress | 2006 | Public benefit forest consists of protection forest and special purpose forest |
| Opinion on Piloting Forest Ecological Benefit Compensation Fund | Financial Ministry and State Forestry Administration | 2001 | Pilot principles, scope, funding, management and measures |
| Measure on Demarcation of Prior Public Benefit Forest | SFA and FM | 2004 | Demarcation principles, detailed scope, standards and application procedure |
| Notice on Demarcation of Public Benefit Forest above the Regional Level | Guangxi Financial Department and Forestry Bureau | 2005 | Operation rules and detailed technical standards for demarcation |
| Notice on Validating the Demarcation Result on Public Benefit Forest | Guangxi Financial Department and Forestry Bureau | 2006 | Validation methods and procedures for public benefit forest |

for demarcation were taken in 1999 and did not reflect the actual situation of forest resources. Currently, the forestry departments are using the updated forest management inventory to adjust the demarcation of the public benefit forests.

However, due to low awareness of local governments on the ecological importance of public benefit forests, Guangxi lagged behind other provinces in making policies for public benefit management. Following the national rules, some pioneering provinces have set up local regulations on management measures for public benefit forests (e.g. Fujian in 2002, Zhejiang and Hainan in 2005, Liaoning in 2006, Inner Mongolia in 2007, Hubei and Jilin in 2008, and Jiangxi in 2009), while the official management measure was still in preparation in Guangxi in 2010. Currently, the management of public benefit forests in Guangxi is mainly based on the Forest Law and its rules for implementation, which cannot provide sufficient institutional support for the management and protection of public benefit forests. China's state authorities often use responsibility contracts to distribute different tasks from high-level to low-level government. In Guangxi, the Letter of Commitment for the Target to Forestry Ecological Protection and Management serves to require the officials at municipal and county level to accomplish the protection and management of the public benefit forests within their jurisdiction. However, the target responsibility system used by governments involves a set of targets in which public benefit forest protection is only one target and regarded as less important than other targets such as GDP. At the same time, there is no corresponding mechanism for rewards and punishments for public benefit forest management at local level. Therefore, local officials have not paid enough attention to public benefit forest protection and management. Furthermore, detailed specifications lack for the establishment of management organization, staff, monitoring, payment etc. Most county forestry departments have not set up a specific office for public benefit forest management. The responsibility of management and protection is usually fragmented among different divisions such as the financial office, the forestry administration office, and the forest resource management office, among which it is hard to effectively coordinate the implementation of the payment schemes. Insufficient governmental budget for forestry departments lies at the root of this problem. The forestry departments at all levels in Guangxi employed 18,787 staff. 55% of them are supported fully by governmental funding; 13% partially supported; and 32% with no governmental funding[24] (Guangxi Public Benefit Forest Protection and Development Task Force, 2009). The forestry departments have not enough financial resources to maintain specific offices for public benefit forest management. As a result, Guangxi lags behind many provinces in making policies and mobilizing financial sources for public benefit forest management.

### 5.2.3 Forestry industrial policy in Guangxi

Guangxi is one of the prestigious fast-growing and high yield plantation bases in China. Its natural condition and a vast forestland are regarded as an advantage to develop plantations for timber production. The plantation area ranks first in China. Some scholars think that the development of plantations pushes for local economic growth, improves the livelihood of people living in mountainous regions, reduces timber harvesting on natural forests and indirectly protects local environment (Wang, 2005). However, the media, environmental NGOs and some scholars express concerns on the negative impacts of plantation development on the local environment, such as biodiversity and underground water (Economy&Nation Weekly, 2010; Greenpeace, 2010). Since the plantation is established within a commercial forest, which is outside the range of the public benefit forest under the classification-base forest management, it is not the focus of this contribution to discuss the impacts of plantations on the environment. However, its policy development has apparent effects on the management of public benefit forests. On the one hand, the development of plantations increases timber supply and thereby to some extent reduces the pressure on public benefit forests. In 2010, Guangxi's commercial forests reached 9.7 million hectare, accounting for 65% of its forestland and its timber production during the 11th Five Year Plan (FYP) is 13.13 million cubic meter, ranking first in China. In general, the development of plantations improves productivity of commercial forests and increases timber production. The overall demand on natural forests has been reduced. On the other hand, the development of plantation policy causes difficulty in the management of public benefit forests. Since 2002, Guangxi government developed a set of policies to support plantation, including subsidy on afforestation, reduction of taxes and fees on timber harvesting, and investment. On the contrary, the payment

---

[24] The deficit is usually balanced by the afforestation fund which the forestry departments levy on timber harvesting for commercial purpose. Since the collective forest tenure reform, most provinces have started to reduce or abolish the afforestation fund.

on the public benefit forest is at an extremely low level. The gap between commercial forests and public benefit forests becomes larger and larger. It drives local farmers to have a negative attitude on public benefit forest management and to show more enthusiasm for developing plantations. In some regions, local farmers even start to request changing their public benefit forest into a commercial forest. There is something missing in Guangxi's current forestry industrial policy. It only focuses on commercial forests without integrating public benefit forests into its plan and it does not take into account the side effects on public benefit forests. In other words, contradictions exist between the development of forest industrial policy and the payment schemes for ecological benefit forest in Guangxi.

## 5.3 Payment schemes in Guangxi Zhuang Autonomous Region

In Guangxi, the central and regional payment schemes determine the field of public benefit forests at central and regional levels. The functions of public benefit forests in Guangxi are described as water source conservation, soil erosion prevention, and nature reserves. Economic functions and recreational functions of the public benefit forest are also allowed as long as the utilization does no harm to the forest ecosystem.

The principles guiding forest utilization in Guangxi include environmental function priority, economic incentive, and administrative responsibility. The priority for ecological services of forest ecosystem demands for a long-term orientation to put ecological conservation in an advantageous position in the utilization of public benefit forests.

The payment schemes also use economic incentives to encourage forest protection instead of traditional "command and control". Although this principle of economic incentives still play a limited role in the payment schemes because of low payment standards, it has been accepted widely. Presently, two types of payment schemes in Guangxi are running to support public benefit forest management and protection. The first one is the central fiscal compensation fund for national public benefit forests. Since 2001, Guangxi has been included as a pilot area for the central compensation fund and the payment for public benefit forests reached 362.9 million Yuan. The second type is the regional payment scheme which has been established in 2006 to support regional public benefit forests[25]. However, Guangxi is a less developed region in China and the regional government has not enough financial resources for the payment scheme. Therefore, it has to rely on extra funding from the central government[26]. In 2009, the central government provided 74.65 million Yuan for regional-level public benefit forests while the regional government shared 30.7% of the payment (31.1 million Yuan) for the regional-level public benefit forests (Table 5.3). In 2010, the payment standard rose to 15,000 Yuan per km$^2$, twice as much as the payment standard before. If it follows this trend, the principle of economic incentive can play a more important role in public benefit forest protection and provide significant momentum to this field.

The third leading principle in the payment schemes is administrative responsibility. The implementation of payment schemes relies on the target responsibility system within the forest

---

[25] It was much later than other provinces and regions and remained at a very low payment level.

[26] In general, the central government provides funding for national public benefit forests and the provincial or regional governments on their own fund the public benefit forests at the provincial or regional level.

*Table 5.3. Funding sources for public benefit forest (PBF) payment schemes in Guangxi, 2001-2009.[1]*

| Year | Central government | | | Provincial government | |
|------|-----------|-----------|-----------|-----------|-----------|
| | area (km²) | payment for national-level PBF (million Yuan) | payment for regional-level PBF (million Yuan) | area (km²) | payment for regional-level PBF (million Yuan) |
| 2001 | 23,333.33 | 175 | 0 | 0 | 0 |
| 2002 | 23,333.33 | 175 | 0 | 0 | 0 |
| 2003 | 23,333.33 | 175 | 0 | 0 | 0 |
| 2004 | 23,333.33 | 172.5 | 0 | 0 | 0 |
| 2005 | 23,333.33 | 172.5 | 0 | 0 | 0 |
| 2006 | 29,066.67 | 218 | 0 | 2,666.67 | 22.4 |
| 2007 | 45,400.00 | 287.85 | 52.65 | 3,000.00 | 22.5 |
| 2008 | 45,400.00 | 287.85 | 52.65 | 4,133.33 | 31.1 |
| 2009 | 48,333.33 | 287.85 | 74.65 | 4,133.33 | 31.1 |

[1] Data provided by the Planning and Financial Division, Forestry Department of Guangxi Zhuang Autonomous Region.

administrative structure. This principle has been to a large extent applied in order to fulfill the objectives of the payment schemes. However, this leading principle also works with other principles such as consultation and local participation. This will be discussed in the evaluation section on participation.

There are several long lasting issues in Guangxi's forest conservation sector, including lack of funding for forest protection, low willingness of establishing and maintaining public benefit forests, and strong human intervention in forest ecosystems. The aims of the schemes are to provide financial support for forest protection, and to facilitate and strengthen forest protection. If these aims can be fulfilled, it is expected that existing public benefit forests can be protected properly, forest degradation can be reduced and ecological services from forests can be maintained. In addition, the payment schemes clarify both the tasks of forestry and financial departments at all administrative levels and how they complete their tasks on payment distribution and use. The schemes also stipulate the division of tasks among various stakeholders, including governments, village committees, foresters, individual forest owners, and local farmers. The payment schemes in Guangxi put emphasis on utilizing the payment for forest management and protection through the forestry administrative apparatus.

## 5.4 Introduction on the case sites

Three state-owned forest farms and two villages have been chosen as representative case sites for evaluation. The state-owned forest farms were chosen based on 3 criteria: farm types, the starting time of payment schemes (the earlier, the better), and forests with high ecological importance. The three state-owned forest farms chosen include two farms mainly for ecological purpose (Pingling and Dayuan) and one for commercial purpose (Huangmian). Table 5.4 shows the main characteristics of the three forest farms.

Pingling Forest Farm locates in Yongfu County and 82% of its forest is used for ecological protection along the Luoqing River, which is an important secondary branch of the Pearl River. The main ecological function of its forest ecosystem is to prevent soil erosion and reduce sediment run off for the hydropower stations. Currently two hydropower stations lie along the Luoqing River and another four stations are in development.

Dayuan Forest Farm is within the jurisdiction of Yangshuo County and manages a sizable and well-reserved natural forest with high ecological importance (Figure 5.2). The forest farm shelters

*Table 5.4. Characteristics of the sampled state-owned forest farms.[1]*

| Farm | Established | Employees | Total area (ha) | Commercial forest (ha) | Public benefit forest (ha) |
|------|-----------|-----------|-----------------|------------------------|----------------------------|
| Huangmian | 1957 | 1,100 | 21,660 | 19,009 | 2,651 |
| Pingling | 1979 | 110 | 9,803 | 1,738 | 8,065 |
| Dayuan | 1958 | 282 | 8,592 | 3,033 | 5,559 |

[1] Source: Investigation on Huangmian, Pingling and Dayuan Forest Farm 2009.

*Figure 5.2. Dayuan Forest Farm, Guangxi.*

a water source and is the rain collecting area for the Dayuan River, which is the main water supply for the River Li, one of China's most famous scenic areas. In addition, the forest guarantees stable water supply for five villages of Xingping Town for daily drinking water and irrigation.

Huangmian Forest Farm also lies along the Luoqing River, but most of its forest is used for commercial purpose. This farm has been chosen to compare with the two ecological type forest farms. In addition, Huangmian Forest Farm is a provincial-level state-owned forest farm and is financially supported by the Guangxi Regional Forestry Department, while Pingling and Dayuan forest farms are county-level state-owned forest farms under the management of the county forestry bureaus.

In Guilin City (including 7 counties within its jurisdiction), which forms the major case site of this research, public benefit forest reaches 5,566.67 km², ranking second in Guangxi. The counties under Guilin City have similar sizes of public benefit forest area, ranging from 680.00 to 833.33 km². Lingchuan County has 740.00 km² of public benefit forest and 86.5% of public benefit forest has been covered by the payment scheme with 5.64 million Yuan every year.

Two villages (Dongyuan Village and Xiling Village, Figure 5.3) in Qingshitan Township of Lingchuan County have been chosen as case study sites to assess the implementation of payment schemes on collective forestland (Table 5.5). The criteria for choosing the county and the villages include the starting time of payment schemes (the earlier, the better), the area of the public benefit forest, forests with high ecological and economic importance, and community's dependence on forests. Lingchuan County proved to be an excellent case study area, fulfilling best these criteria. Lingchuan County lies northeast of Guilin City, one of the famous tourism cities in China. Its jurisdiction includes 11 townships, 2,258 km² and a population of 355,000. The county is located in mountainous region with forest coverage of 65.9%. The public benefit forest (52.8% of the forestland in the county) provides two major ecological services: watershed protection and landscape value preservation for several forest tourism sites. The two sampled villages are both located in the water source protection area. Table 5.5 shows the characteristics of the sampled villages. 15 farmer households from each village have been chosen according to differences in age, education and family size. In total, 30 farmers attended the group meeting and received the questionnaires.

*Table 5.5. Characteristics of the sampled villages.[1]*

| Village | Land area (ha) | Households | Population | income per capita (Yuan) | Public benefit forest (ha) |
|---------|----------------|------------|------------|--------------------------|----------------------------|
| Dongyuan | 5,945 | 380 | 1,510 | 3,000 | 5,673 |
| Xiling | 4,181 | 500 | 2,080 | 3,000 | 4,040 |

[1] Source: Qingshitan Township Forestry Station 2009.

*Figure 5.3. Xiling Village, Lingchuan County, Guangxi.*

## 5.5 Implementation of survey

Interviews have been carried out in the forestry departments at provincial, municipal and country level and at the State-Owned Forest Farms (SOFFs) in November 2009. The aim of the interviews was to understand the process of demarcation, management and examination of the public benefit forest in practice, to understand the use of funds from the government and other sources, and to assess their attitudes to and expectations on the schemes. Through the interviews, a lot of material has been collected. In addition, a questionnaire has been used on the managers of the three state-owned forest farms to collect data on forest resources and forest management relating to public benefit forest.

Two group meetings including village leaders and local farmers have been carried out in two villages. The farmers were selected with the help of village heads to represent different household economic conditions (high, middle and low income) and livelihood dependency on forests. The group meeting elicited major concerns of local farmers on the payment policies. Through the group meeting, the farmers discussed how they managed their public benefit forest, what they thought about the payment schemes, and what impact the schemes had on their income and daily life. The discussion of the group meeting showed that the farmers had a general opinion on the payment schemes and there was no big division among them. Two issues have been raised by most of the farmers: insufficient payment from governments and the lack of forestland to develop timber and bamboo forests.

After the group meeting, questionnaires have been sent out to the farmers in order to obtain detailed information. The questionnaires asked farmers about their income change, participation in each stage of the implementation of the payment scheme, attitudes toward payment regulations and impression on the performance. 19 out of 30 questionnaires distributed have returned. The

*Table 5.6. Characteristics of farmer households of Lingchuan County.[1]*

| Gender | male | | female | | |
|---|---|---|---|---|---|
| (head of household) | 19 | | 0 | | |
| **Ethnic group** | Han | | minority | | |
| | 19 | | 0 | | |
| **Age** | 30-39 | 40-49 | 50-59 | ≥60 | |
| (head of household) | 6 | 8 | 2 | 3 | |
| **Education** | Senior | | Junior | | primary |
| (head of household) | 7 | | 9 | | 3 |
| **Family size** | ≤3 | | 4-6 | | ≥7 |
| | 4 | | 11 | | 4 |
| **Children** | ≤1 | | 2 | | ≥3 |
| | 10 | | 7 | | 2 |

[1] Source: farmer household survey in Lingchuan County in November 2009.

farmers were active in expressing their opinion on the payment schemes during the group meeting but some of them were reluctant to fill in questionnaires individually. Firstly, they might feel safer to express their dissatisfaction on the schemes in a group, rather than by individual. Furthermore, detailed data on their income and forests required for the questionnaires iced on their enthusiasm and some farmers thought that the group meeting was enough to express their opinion and it was unnecessary to do it again in questionnaires. Since the questionnaires returned shows similar results as the group meeting, there is not significant bias in the eleven farmers that chose not to fill out the questionnaires. Table 5.6 shows the characteristics of farmer households from the villages. All respondents, the heads of the households, were male and Han people with ages from 30 to 68 years old. The age structure of respondents is slightly younger than the population of Guangxi. All of them completed primary education and most of them finished middle school education. Their education levels are higher than the average levels of rural population in Guangxi, which has 15% finishing senior high school and 60% with junior high school education in 2008. Younger and relatively educated respondents could imply more alternative employment opportunity and stronger capacity to adapt to the change of household economy caused by the payment schemes. The family size of the sampled households ranges from 3-8 members including 0 to 4 children (younger than 18 years old). The average per capita income in the sampled villages is 3,000 RMB, which is lower than the average of rural residents in Guangxi in 2008 (3,690 RMB). It is common for forested regions, where farmers are more seriously stricken by poverty than other rural areas.

## 5.6 Evaluation on the implementation of payment policy by state-owned forest farms

The analysis on the performance of the payment schemes includes three parts: effectiveness, economic impact and participation mechanism.

In order to evaluate the performance of the payment schemes, five plots of public benefit forests have been selected for both Pingling Forest Farm and Dayuan Forest Farm, according to the main characteristics of the public benefit forest of the forest farms. Selection criteria included forest quality (at different levels), forest types (natural forest and plantation), and ecological importance. Huangmian Forest Farm provided aggregated data on all of its public benefit forests since its forest plots can be categorized into four types and each type has a similar resource situation.

### 5.6.1 Environmental effectiveness

Forest resource data have been collected from each forest farm. Table 5.7 shows the situation of forest resources before the payment schemes[27]. About half of the forest plots in two forest farms (Pingling and Dayuan) were plantations for timber production before the payment schemes. The stock volume, which is an important indicator for measuring the quality and the capacity of forest to provide ecosystem services, varies to a large extent among forest farms. On average, Dayuan Forest Farm has the highest stock volume, reaching around 10 cubic meters, while the forests in Pingling Forest Farm only have a stock volume of around 3 cubic meters. However, the design of payment schemes neglected the difference of the original forest quality completely. As a result, a uniform payment standard has been applied to all forest plots and forest owners with higher average stock volume suffered from greater economic loss.

Although the government requires the demarcation to comply with standards relating to ecological importance, there is still some space for the forest farms to select less economically profitable plots into the public benefit forests. Generally, the forest farms would like to keep forest plots for timber production on a low slope with good soil quality and less distance to roads. The sample plots show that most of public benefit forests in the 3 forest farms are located on land with over 25-degree slopes, with average soil quality and over 5 kilometers from roads.

According to the interviews, the managers of the forest farms indicated that they had taken economic development of the forest farms into consideration when demarcating and adjusting the public benefit forest. For example, first Huangmian Forest Farm demarcated 46.67 km² of forests for public benefit purpose in 2001. However, after they decided to develop as a commercial type

---

[27] The earliest public benefit forest demarcation started in 1999. All forest plots in Pingling and Dayuan Forest Farm have been demarcated as public benefit forest in 1999. The forest plots in Huangmian Forest Farm have been added into public benefit forests in different periods from 2000 to 2004.

*Table 5.7. Situation of forest resources before the payment schemes.*

| Plots[1] | Area (ha) | Year | Main usage | Volume[2] | Type | Generation | Age | Slope | Soil quality | Distance to road[4] |
|---|---|---|---|---|---|---|---|---|---|---|
| PL1 | 4.7 | 1999 | timber | 2.2 | *Pinus massoniana* | plantation | 12 | 20° | average | 1 |
| PL2 | 6.3 | 1999 | timber | 2.6 | *P. massoniana* | plantation | 10 | 28° | average | 7 |
| PL3 | 7.0 | 1999 | timber | 3.7 | *Cunninghamia lanceolata* | plantation | 8 | 15° | good | 4 |
| PL4 | 4.2 | 1999 | timber | 3.5 | broad-leaved | natural | 40 | 35° | average | 8 |
| PL5 | 6.7 | 1999 | timber | 3.6 | broad-leaved | natural | 20 | 28° | average | 10 |
| DY1 | 11.4 | 1999 | timber | 14.0 | *C. lanceolata* | plantation | 28 | 30° | average | 1 |
| DY2 | 35.8 | 1999 | timber | 12.0 | *C. lanceolata* | plantation | 26 | 30° | average | 1.5 |
| DY3 | 84.7 | 1999 | protection | 10.0 | broad-leaved | natural | 30 | 35° | average | 2.5 |
| DY4 | 49.2 | 1999 | protection | 10.0 | broad-leaved | natural | 40 | 35° | average | 8 |
| DY5 | 108.1 | 1999 | protection | 8.8 | broad-leaved | natural | 30 | 35° | average | 5 |
| HM1 | 247.2 | 2004 | protection | –[3] | brush | natural | 0-15 | <25° | average | 10 |
| HM2 | 804.2 | 2000 | protection | –[3] | brush | natural | 0-20 | ≥25° | average | 10 |
| HM3 | 321.7 | 2001 | protection | 4.0 | broad-leaved | natural | 2-45 | <25° | good | 5 |
| HM4 | 1,277.9 | 2000 | protection | 5.0 | broad-leaved | natural | 2-42 | ≥25° | good | 5 |

[1] PL1-5 are the forest plots selected in the Pingling Forest Farm; DY1-5 for the Dayuan Forest Farm; HM1-4 are the aggregated forest areas of the Huangmian Forest Farm.
[2] Volumes are the average stock volume (m³) per 0.067 ha on each forest plots.
[3] No data are available for the stock volume of brush in Huangmian Forest Farm.
[4] Distance to road indicates the distance of the forest plots from roads available for vehicles, in km.

forest farm, the public benefit forest has been reduced to 26.67 km². In the present, they plan to add 13.33 km² of forests to the public benefit forest, which they thought had low economic value and high ecological importance. Similarly, Dayuan Forest Farm also turned some public benefit forest into commercial forest between 2005 and 2008. This flexibility of selecting forest plots for state-owned forest farms brings both advantages and disadvantages. The advantage is that the forest farms can arrange the forest plots in a more efficient way to enhance their economic benefit such as using forests at good locations for timber production and integrating public benefit forest plots into a whole area, which is easier for management, protection and benefits biodiversity. However, frequent adjustment and too much emphasis on the economic benefit may cause reduced environmental effectiveness of the public benefit forest. Some plots with both important economic and environmental importance will more likely be used for timber production than for ecological restoration.

Table 5.8 shows the changes of forest resources from 1999 to 2008, in terms of usage, tree species and stock volume. Because a severe snow disaster in 2008 imposed huge damage to the forest ecosystem, the whole period is divided into two segments. The first segment from 1999 to 2007 showed the effect of the payment schemes on the quality and structure of public benefit forest. The last period from 2007 to 2008 demonstrated the impact of the snow storm in 2008, which also reflected some problems in the structure of the public benefit forest. Since the payment schemes started, all forest plots have changed into public benefit forests for water source protection and/ or soil erosion prevention. The tree species on all forest plots stayed the same as before, except for two plots of Dayuan Forest Farm, on which *Cunninghamia lanceolata* has been harvested for timber and replanted with *P. massoniana*. From 1999 to 2007, the stock volumes of both Pingling and Dayuan forest farm's forest plots have increased. Comparably, the plantations have higher growth rates than natural forests and are more than twice their stock volume. The forests (PL4, PL5), which were used for timber production before the payment schemes have grown faster than old protection forests (DY3-5). First, the timber forests started from lower stock volumes with around 3.5 cubic meters, compared with around 10 cubic meters of the protection forests. Second, the protection forests before the payment schemes have entered into a stable and mature system. This can also be proven by the fact that the plots DY3-5 have a relatively lower loss during

*Table 5.8. Change of forest resources from 1999 to 2008.*

| Plots | Main usage[1] | Type | Age | Volume[2] in 2007 | Vol. change rate (%) | Volume[2] in 2008 | Vol. damage rate (%) |
|-------|-----------|------|-----|------------------|---------------------|------------------|---------------------|
| PL1 | protection | *Pinus massoniana* | 21 | 5.1 | 133 | 1.8 | 65 |
| PL2 | protection | *P. massoniana* | 19 | 5.2 | 104 | 1.5 | 72 |
| PL3 | protection | *Cunninghamia lanceolata* | 17 | 8.6 | 130 | 4.3 | 50 |
| PL4 | protection | broad-leaved | 49 | 5.0 | 40 | 2.3 | 53 |
| PL5 | protection | broad-leaved | 29 | 6.4 | 80 | 1.9 | 70 |
| DY1 | protection | *P. massoniana* | 8 | 3.0 | – | 1.5 | 50 |
| DY2 | protection | *P. massoniana* | 8 | 3.0 | – | 1.5 | 50 |
| DY3 | protection | broad-leaved | 39 | 12.0 | 20 | 8.0 | 33 |
| DY4 | protection | broad-leaved | 49 | 12.0 | 20 | 8.0 | 33 |
| DY5 | protection | broad-leaved | 39 | 10.0 | 14 | 8.0 | 20 |
| HM1 | protection | brush | 0-19 | –[3] | – | – | – |
| HM2 | protection | brush | 0-28 | –[3] | – | – | – |
| HM3 | protection | broad-leaved | 2-52 | –[4] | – | 5.0 | – |
| HM4 | protection | broad-leaved | 2-50 | –[4] | – | 5.0 | – |

[1] The shadowed cells indicate the plots which have changed in main usage or tree species.
[2] Volumes are the average stock volume (m$^3$) per 0.067 ha on each forest plots.
[3] No data are available for the stock volume of brush in Huangmian Forest Farm.
[4] No data are available for the stock volume of public benefit forest in Huangmian Forest Farm in 2007.

the snow disaster in 2008. In conclusion, the payment schemes changed the usage of the forests and the quality of public benefit forest has been improved since it started.

In early 2008, an unusually severe snow storm occurred over southern China. A total of 18.6 million hectares of forests, about one-tenth of China's forest resources, have been damaged by the unprecedented snow wreckage in at least five decades, with forests, bamboo and seedlings seriously destroyed in 19 provinces including Guangxi, Hunan, Guizhou, Jiangxi, Hubei, Anhui, and Zhejiang. In Guangxi Zhuang Autonomous Region, Guilin (including the 3 forest farms and 2 villages selected in this research) was among the worst-hit regions, with 678.2 thousand hectares of forests damaged, including 347.7 thousand hectares of pine and *Cunninghamia*, 12.0 thousand hectares of *Eucalyptus*, 93.0 thousand hectares of bamboo, and 225.5 thousand hectares of other tree species (Hao and Ou, 2008). The value of related losses reached 3.7 billion Yuan RMB.

Besides the loss of timber forest, the snow disaster badly damaged the local forest ecosystem. According to the estimate of the survey by the SFA, forest coverage in Guangxi decreased 3-4% and in Guilin, coverage went down by more than 10%. As a result, the ecological function of this region for protecting the water source of the River Li has been badly impacted. Table 5.8 indicates that the stock volumes have fallen sharply from 2007 to 2008. In the Pingling and Dayuan forest farms, the stock volumes of the public benefit forest have been reduced by 20%-70% and became lower than the levels in 1999. For Huangmian Forest Farm, although there are no data available on volume change, the survey of the SFA shows that a large area of its forests at an altitude above 300 meters has been ruined, including 6.67 km$^2$ of pine, Cunninghamia, and broadleaf trees, accounting for one-fourth of its public benefit forest.

Five indicators including biodiversity, canopy density, phytocenosis structure, vegetation coverage and depth of litter layer have been chosen to measure the change in ecological functions of the public benefit forest. Table 5.9 shows the change of ecological indicators for the public benefit forest in Pingling Forest Farm from 1999 to 2007. The Pingling Forest Farm demarcated its public benefit forest in 2001. Until 2007, the natural forest plots (PL4 and 5) have been maintained as broadleaf dominated forest and the original plantation plots (PL1-3) have evolved into mixed forest with over 30% of broadleaf trees. The canopy density of all public benefit forest plots also increased to around 0.8 in 2007 and a similar expansion also happened to the shrub and grass coverage. The phytocenosis structure of all public benefit forests has been improved to a complicated level with more than two-tier trees, shrub and grass until 2007. At the same time, the litter layer on the forest has doubled, which makes it better to block raindrops. However, the snow storm in early 2008 wrecked the forest ecosystem and reduced the canopy density and the coverage of shrub and grass on the public benefit forests. The canopy density reduced to a level lower than in 1999 and the coverage of shrub and grass in general returned to the level in 1999. The litter layer was not affected.

In conclusion, the payment schemes changed forest use practices of forest farms, improved the quality of the forest and enhanced its capacity for providing ecosystem services. The devastating snow storm in 2008 caused vast damage to public benefit forest. Therefore, forest management of the payment schemes should put particular emphasis on optimizing the structure of the forests and strengthening resilience of forest ecosystem in the future.

*Table 5.9. Ecological indicators[1] of public benefit forest plots in 1999 (before payment schemes), 2007 and 2008 of Pingling Forest Farm.*

| Plot | Year | Biodiversity | Canopy density | Phytocenosis structure | Shrub and grass coverage | Litter layer (cm) |
|------|------|-------------|----------------|------------------------|--------------------------|-------------------|
| PL1 | before | 3 | 0.5 | 3 | 0.5 | 2 |
|     | 2007 | 2 | 0.8 | 1 | 0.8 | 5 |
|     | 2008 | 2 | 0.4 | 1 | 0.4 | 5 |
| PL2 | before | 3 | 0.3 | 2 | 0.5 | 2 |
|     | 2007 | 2 | 0.7 | 1 | 0.7 | 5 |
|     | 2008 | 2 | 0.3 | 1 | 0.5 | 5 |
| PL3 | before | 3 | 0.6 | 2 | 0.5 | 2 |
|     | 2007 | 2 | 0.8 | 1 | 0.7 | 4 |
|     | 2008 | 2 | 0.5 | 2 | 0.5 | 4 |
| PL4 | before | 1 | 0.7 | 1 | 0.5 | 3 |
|     | 2007 | 1 | 0.8 | 1 | 0.75 | 5 |
|     | 2008 | 1 | 0.7 | 1 | 0.5 | 5 |
| PL5 | before | 1 | 0.7 | 2 | 0.5 | 2 |
|     | 2007 | 1 | 0.8 | 1 | 0.75 | 4 |
|     | 2008 | 1 | 0.6 | 1 | 0.5 | 4 |

[1] The detailed description on indicators is provided in Chapter 2.

## 5.6.2 Cost benefit analysis and efficiency

This part deals with costs and benefits relating to the payment schemes. The cost-benefit analysis has been carried out on two levels (individual farms and the forest plots). First, expenditures and payments related to the management of public benefit forest are analyzed to evaluate the economic impact of the payment schemes. Then, a simplified cost-benefit analysis has been carried out, based on economic costs and benefits data from each forest plot, to explore the impacts on different types of forest plots.

The benefits of the forests in this area are timber production, recreation, and ecosystem services (soil erosion prevention, water source protection, carbon sequestration, biodiversity, etc.). Although forests also have existence values and bequest values, these values are very difficult to measure in economic terms. On the cost side plantation, management, and cutting (selective or clear) are the most obvious costs. In addition, opportunity costs should be considered, since after the payment schemes, the forest farms have to stop timber production for public benefit forest. Then net revenue from timber production will be the opportunity cost of the forest farms.

## Financial situation of the state-owned forest farms and the payment

This part introduces the whole financial situation of the state-owned forest farms, their expenditure on public benefit forest management, and the financial burden on them. The three forest farms have quite different financial situations, as became apparent through the interviews with their directors. Huangmian as a provincial state-owned forest farm has the best economic condition and diverse income sources. After the overall downturn of state-owned forest farms, Huangmian Forest Farm started to adjust its tree species and to develop fast-growing and high yielding plantations in 2002. Meanwhile, it succeeded in attracting investment for developing forest industry including setting up a tea factory, a medium density fiber board factory and an artificial board factory. Under such a good economic condition, management of a small area of public benefit forest is a relative low burden on the farm, although there is about 70,000 Yuan of annual deficit.

Dayuan Forest Farm is under the jurisdiction of Yangshuo County, but actually with respect to finance, it depends on its own, since the county is too poor to provide sufficient budget for the forest farm (only 30,000 Yuan RMB is allocated as annual subsidy). Dayuan forest farm's economic condition ranks in the middle, with two major income sources: timber harvesting and forest-related tourism. The farm kept a sizeable commercial forest after the demarcation of public benefit forest. Most of its commercial forest has been planted between 1958 and 1982. After 27 years of logging, the stock volume of its near-mature forest is less than 20,000 cubic meters, which will be cut in only 3 years if the current timber production continues. The forest farm also tried to replant pine and *Cunninghamia* trees but the replanting has been a failure because of high altitude (over 600 meters above sea level) and single tree species. If no suitable tree species are replanted, the forest farm will be out of timber forest for income in 3 years. This part of income is very important for the forest farm to complement its deficit in management of the public benefit forest. In 2005, the Yangshuo National Forest Park was set up under the management of the Dayuan Forest Farm. The park also contributed its forest resources to become a shareholder of some recreation projects such as river drifting. Currently, the forest farm can get an annual share of 240,000 Yuan from recreation projects. However, the income from tourism alone cannot offset the deficit from the public benefit forest.

Pingling Forest Farm has the worst economic situation, with little commercial forest resources and no other income sources. Its revenue mainly depended on the payment for public benefit forests. It has a deficit of 830,000 Yuan for management of its public benefit forest in 2008. Table 5.10 shows the incomes and the expenditures which each farm received and spent in 2008. The income is just enough for the salary of foresters and the forest farms had to find other funds for forest tending, road maintenance and fire control. The deficit in funding decreased the incentive and the capacity of the state-owned forest farms in public benefit forest management and protection and they have to engage in development projects (timber forest) to meet the financial gap.

## Cost benefit analysis for public benefit forest plots

In order to understand the gap between costs and benefits and its change on different forest plots after the payment schemes were launched, we collected all cost and benefit data of the selected forest plots from the establishment of forestland to 2008 and the monetary value of all prices

*Table 5.10. Payment of the schemes and expenditure for public benefit forests in 2008 (10,000 Yuan).[1]*

| Forest farms | Payments[2] | Expenditure | | | | | | Deficit |
|---|---|---|---|---|---|---|---|---|
| | | total | salary | forest tending | road maintenance | fire control | other | |
| Pingling | 200 | 283 | 200 | 40 | 29 | 8 | 6 | 83 |
| Dayuan | 38 | 79 | 40 | 13 | 1 | 9 | 16 | 41 |
| Huangmian | 18 | 24 | 20 | 0 | 0 | 4 | 0 | 6 |

[1] Interviews with the managers of the 3 forest farms.
[2] The payments refer to annual payments from the central and provincial payment schemes for forest ecological services.

has been adjusted to real terms, taking 2008 as the base year (Table 5.11). The result showed that before the payment scheme Pingling Forest Farm did not obtain any revenue from its plantation since its plantation plots were still in middle age and not available for final logging. Pingling Forest Farm has carried out selective logging for two plantation plots (PL1 and PL3) in order to improve forest growth. Although some timber from such logging can be sold, the revenue of the timber generally cannot offset the costs of selective logging. So before the payment schemes, Pingling Forest Farm already has invested 240.0-403.5 Yuan RMB per ha in its plantation each year. For natural forest plots, the forest farm took them as reserved land for plantation and spent 82.5-145.5 Yuan RMB per ha for management annually. Dayuan Forest Farm had finished a round of final logging[28] on its plantation (DY1 and 2) before the scheme was launched. The analysis showed it had a net benefit of about 38 Yuan RMB on its plantation plots on average each year. It is also an apparent example to demonstrate the opportunity cost of the land for protection. Dayuan Forest Farm kept relative low management costs on its natural forest plots (DY3-5), less than 15 RMB per ha every year. The natural forest plots of Dayuan had been indicated as protection forest by the government before the payment schemes and it was the obligation of the forest farm to take care of the protection forest, even without financial support from the government. Huangmian Forest Farm had a medium cost level for the management of its protection forest and brush, around 45 Yuan RMB per ha each year.

Table 5.12 shows costs and benefits of public benefit forest management after setting up the schemes and also the changes in net benefit of different types of forest plots. Specifically, the "net change" indicates the difference of net benefit generated from each plot between before (Table 5.11) and after payment schemes (Table 5.12). Under the schemes, all final cutting has stopped for public benefit forests. For Pingling Forest Farm, it meant that all investment on the plantation plots was lost, without compensation. This change in forest usage turned the plantation plots of

---

[28] Final logging indicates felling down all trees on the plots. After the final logging, forest farms will start a new round of afforestation.

*Table 5.11. Cost and benefit of forest management before the payment schemes (Yuan RMB).[a]*

| Plot | Years | Costs[b] | | | | | Benefits[b] timber | Net benefit[b] | Average net benefit (Yuan/year.ha) |
|---|---|---|---|---|---|---|---|---|---|
| | | afforestation | management | selective harvest | final harvest | total cost | | | |
| PL1 | 13 | 4,760 | 9,246 | 9,172[c] | 0 | 23,178 | 8,400 | -14,778 | -243 |
| PL2 | 11 | 1,690 | 14,900 | 0 | 0 | 16,590 | 0 | -16,590 | -240 |
| PL3 | 9 | 8,295 | 16,519 | 11,714[c] | 0 | 36,528 | 11,130 | -25,398 | -403.5 |
| PL4 | 21 | 0 | 7,303 | 0 | 0 | 7,303 | 0 | -7,303 | -82.5 |
| PL5 | 21 | 0 | 20,307 | 0 | 0 | 20,307 | 0 | -20,307 | -145.5 |
| DY1 | 29 | 10,260 | 66,096 | 0 | 256,500[c] | 332,856 | 525,000 | 192,144 | 580.5 |
| DY2 | 27 | 32,220 | 200,160 | 0 | 734,616[c] | 966,996 | 1,503,600 | 536,604 | 555 |
| DY3 | 31 | 0 | 26,784 | 0 | 0 | 26,784 | 0 | -26,784 | -10.5 |
| DY4 | 36 | 0 | 15,552 | 0 | 0 | 15,552 | 0 | -15,552 | -9 |
| DY5 | 31 | 0 | 40,176 | 0 | 0 | 40,176 | 0 | -40,176 | -12 |
| HM total | 44 | 0 | 5,440,882 | 0 | 0 | 5,440,882 | 0 | -5,440,882 | -46.5 |

[a] The table is calculated from the data provided by the forest farms.

[b] The costs and benefits of forest management on each plot are in total amount from the year of afforestation until 2000.

[c] Selective logging on PL1 and PL3 was carried out in 2000; final logging on DY1 and DY2 was conducted in 1999.

Dayuan Forest Farm (DY1-2) from net gainers into net losers (around 3,000 Yuan per ha each year). Although the payment increased the income of the forest farms, the rapid rise of the cost for forest management, fire control and road maintenance made the gap between cost and benefit on the forest plots wider. In addition, the change of cost and benefit on PL2 and DY1-2 shows that reforestation and selective logging, which is a measure to restore and improve public benefit forest, also increased the cost of the forest farms.

Based on the data above, if the net benefits are taken as a yardstick for measuring the payment, the ideal payment should at least balance the cost and benefit (a zero net benefit). The current payment standard does not offer enough funding for forest management on all forest plots. The additional subsidy for plantation reforestation after the payment schemes should be about 2,400

*Table 5.12. Costs and benefits of public benefit forest management from 2001 to 2008 (Yuan RMB).[1]*

| Plot | Years | Cost | | | | Benefit | | Net benefit | Average net benefit (Yuan/ year.ha) | Net change (Yuan/ year.ha) |
|---|---|---|---|---|---|---|---|---|---|---|
| | | afforestation | management | selective harvest | total cost | timber | payment | | | |
| PL1 | 8 | 0 | 10,388 | 0 | 10,388 | 0 | 2,660 | -7,728 | -207.0 | 36.0 |
| PL2 | 8 | 0 | 18,800 | 15,143 | 33,943 | 10,640 | 3,572 | -19,731 | -393.0 | -153.0 |
| PL3 | 8 | 0 | 21,008 | 0 | 21,008 | 0 | 3,990 | -17,018 | -304.5 | 99.0 |
| PL4 | 8 | 0 | 8,272 | 0 | 8,272 | 0 | 2,394 | -5,878 | -175.5 | -91.5 |
| PL5 | 8 | 0 | 20,000 | 0 | 20,000 | 0 | 3,800 | -16,200 | -304.5 | -159.0 |
| DY1 | 8 | 47,196 | 188,850 | 0 | 236,046 | 0 | 6,498 | -229,548 | -2,517.0 | -3,097.5 |
| DY2 | 8 | 148,212 | 560,550 | 0 | 708,762 | 0 | 20,406 | -688,356 | -2,403.0 | -2,958.0 |
| DY3 | 8 | 0 | 105,000 | 0 | 105,000 | 0 | 48,279 | -56,721 | -84.0 | -73.5 |
| DY4 | 8 | 0 | 52,500 | 0 | 52,500 | 0 | 28,044 | -24,456 | -61.5 | -54.0 |
| DY5 | 8 | 0 | 157,500 | 0 | 157,500 | 0 | 61,617 | -95,883 | -111.0 | -99.0 |
| HM total | 8 | 0 | 1,962,728 | 0 | 1,962,728 | 0 | 1,431,540 | -531,188 | -25.5 | 21.0 |

[1] Calculated from the data provided by the forest farms.

Yuan RMB per ha each year within the first 5 years[29]. For existing plantations, the gap in the payment is 207-393 Yuan RMB per ha each year without consideration of the sunk cost of the forest farm. For natural forests, the additional payment should be 25.5-304.5 Yuan per ha annually, depending on its location and the difficulty for forest management and fire control. Although the payment schemes are the most important financial support for forest farms to protect their public benefit forest, it is still far from sufficient to offset the loss of the forest farms. In addition, the payment standard should consider the rise of all kinds of costs with time.

In order to cope with the losses, these forest farms attempted to generate new income from their forest resources and labor. First, the forest farms with few commercial forestland started to rent land from nearby villages to develop fast-growing and high yielding plantation or reorganize their employees into afforestation team to work for villages. Second, those with unique landscape are seeking investors for developing ecotourism and encouraging their employees to set up small business around touristic spots.

---

[29] The average period for afforestation is 5 years, in which costs are higher than ranging and management afterwards.

### 5.6.3 Participation

The state-owned forest farms have the task of public forest management as local agents of the state forest sector. Their participation in the payment schemes mainly followed the administrative way, which means that the provincial and local forestry departments represent these farms in discussing the public benefit forest and payment issues. The implementation of payment schemes is a political task for the state-owned forest farms, which they have to achieve even without enough funding.

In the implementation of the payment schemes, the state-owned forest farms as local agents, directly carried out demarcation and public benefit forest management. In this sense, the state-owned forest farms had greater power than villages in deciding how to demarcate public benefit forests, how to organize forest management and how to use the payment for public benefit forests.

Different from villages, the state-owned forest farms in Guangxi played an active role in changing forestry production in a sustainable way and reducing the financial deficit in public benefit forest management. Developing forest-related tourism has been often brought up by the managers of Pingling and Dayuan forest farms, when they introduced future development plan. This is also one of the most feasible ways to marketize ecological services of public benefit forest by themselves. However, it is difficult for the forest farms to attract investment from outside, due to their disadvantaged market status, competitive tourism and lack of market information and experience. Pingling Forest Farm is still struggling to seek suitable investments for developing tourism for its forest resources. With convenient accessibility and a unique landscape, Dayuan Forest Farm has succeeded in mobilizing investment in developing tourism. It made a 30 year contract with private investors for using its forest resources to develop tourism (Figure 5.4). Currently, the tourism spot attracts 50,000 tourists and creates 5 million Yuan of gross income every year. However, the forest farm gets only 240 thousand Yuan for using its forest resource from this tourism project each year. The major share of the revenues has been grasped by outside investors.

At present, only the central and provincial governments provide payments for the forest farms to manage and protect the public benefit forest. No funding is available from the municipal and county governments or other beneficiaries such as hydropower stations and tourism companies. The important reason for a lack of local PES initiatives is that there is no formal law or regional

*Figure 5.4. Tourism spots around public benefit forest developed by state-owned forest farms.*

regulation to entitle the forest farms or the governments to collect payments from the beneficiaries of forest ecological services. In addition, the municipal and county governments, where the public benefit forest is located, are generally less developed. Therefore, local governments not only have no capacity to provide more funding for PES but are also reluctant to make a local regulation for PES. The forestry department of the autonomous region has put forward some suggestions for the regional government to increase funding for PES. Four funding channels were suggested: 25% of water resource fee at regional level; 25% of afforestation and reforestation fee at regional level; some part of water and electricity bills for all consumers; and some parts from the revenue of shipping industry and tourism. However, many stakeholders such as the Commission of Development and Reform, the Water Conservation Department, the Environmental Protection Bureau, and the Tourism Bureau, which have been involved in these suggested payment schemes, usually have more power than the Forestry Bureau to influence policies of the autonomous regional government, according to their importance in the administrative system[30]. Therefore, without support of these departments, it is impossible for the Forestry Bureau to implement these funding suggestions via a formalized policy.

## 5.7 Evaluation of the implementation of payment policy by villages

The evaluation of payment schemes implemented by villages includes three parts: environmental effectiveness, livelihood impact assessment and participation mechanism analysis.

### 5.7.1 Environmental effectiveness

Unlike the forest farms, it is difficult to have a quantitative evaluation on environmental effectiveness of payment schemes for villages because the data on forest quality and environmental indicators at the village level are not available in Guangxi. In this part, we use information from the farmer household survey to evaluate environmental performance of the payment schemes in rural areas. Qingshitan Township is located at the northern part of Lingchuan County and has been included into the Guibei (North of Guilin) Prior Protection Area for Water Source and Biodiversity. This protection area, which covers 19,000 square kilometers, provides water conservation services for 8 rivers, including the famous River Li, and protects subtropical evergreen broadleaf forests and rare animal species. The Ecological Function Zoning Plan of Guangxi (Guangxi Regional Government, 2008) identified 4 ecological and environmental risks in this area. First, natural broadleaf forest is reduced and replaced by single species plantation for timber or fruits and such a decreasing quality of forest weakens its function of water conservation. From 1986 to 2002, the area of the forest for water conservation was reduced by 22.8% and especially the proportion of natural forest, which has a strong capacity for water conservation, decreased from 24.68% to 4.12% (Yang, 2005). As a result, the water volume of rivers falls down sharply during dry seasons. The annual low water level period extended from 4 months in the 1940s to 6 months in the 1990s (Wang and Huang, 2008). Second, floods, mudslides and landslides often occur in some parts of this region during

---

[30] Unlike the Fujian Provincial Forestry Department, Guangxi Autonomous Regional Forestry Bureau is at a lower position and had no major say in the regional government until 2010.

rainy seasons. Third, the biodiversity is under serious damage. Fourth, the rivers suffer severe pollution from sewage and waste from urban regions. The first three problems are closely related to forest ecosystem services (water conservation, soil erosion prevention, and biodiversity), most of which the public benefit forest aims to provide.

According to the group meeting and the farmer household survey, most farmers thought their village had some soil erosion before the payment schemes started. However, about one-third of the farmers interviewed believed their villages had no apparent soil erosion. Therefore, soil erosion was not widely spread in the villages and only impacted some parts of the region before the payment schemes. Around 60% of all respondents thought that there was no change in the local environmental qualities following the payment schemes, while 30% of them believed that the soil erosion has been reduced slightly. Therefore, from the perspective of local farmers the payment schemes played a role of maintaining the local environment rather than improving it. Through the group meeting with local farmers, some factors have been identified to have negative impacts on the environmental effectiveness of the payment schemes. First, the farmers increasingly replaced natural broadleaf stands with bamboo stands, since bamboo logging is an important profitable income source of local communities and not subject to logging permission from the government. In 1999, the proportion of bamboo forest in the land along the upper reach of the River Li was only 7.7% on average (Lu *et al.*, 2002). The bamboo proportion of Dongyuan and Xiling Village reached to 18.3% and 31.6% respectively in 2008. However, the contribution of bamboo forest to water conservation is less effective than that of natural evergreen broadleaf forest (Gao, 2006; Lu *et al.*, 2002; Sun *et al.*, 2008). In addition, the production activities including digging shoots, reclamation, logging and skidding deteriorated soil erosion in the region, as the farmers generally did not have the knowledge about soil erosion or did not take any measure to prevent it (Gao, 2006). Second, the farmers illegally harvested the public benefit forest and harvested the forests more than allowed by the permission for selective logging. According to the interviews with the heads of the villages, 5% of the farmer households, which were generally stricken by poverty, have carried out illegal logging after the payment schemes. Although the Forestry Bureau required strict standards on selective logging, it is very difficult for the bureau to monitor the logging practices of local farmers. Furthermore, if the farmers stick to the selective logging requirements, the cost of logging will increase sharply, even beyond the revenue. In order to strengthen management and protection, the Forestry Bureau sometimes stopped approval for selective logging on public benefit forest.

### 5.7.2 Cost benefit analysis and livelihood impact

The county forestry bureau included almost all forest land of the two villages into the public benefit forest in 2001. The implementation of payment schemes reshaped the way of work and life in villages. According to the survey, the average household income decreased by 36.6% from 15,453 Yuan in 2001 to 9,791[31] Yuan in 2007. Before the payment schemes started, the economy of the

---

[31] The monetary value of farmer household income in 2007 have been adjusted into real terms (taking 2001 as the base year), according to the consumption price index for rural residents provided by China Rural Statistics Yearbook 2010 (National Bureau of Statistics, 2010).

villages heavily relied on timber and bamboo logging. 68.4% of farmer households took forestry production as the main source of household income and on average, timber and bamboo logging accounted for 60.0% of household income (Figure 5.5). After the introduction of the payment schemes, off-farm work replaced timber logging as the main income source. Only 31.6% of the households still mainly depended on timber and bamboo logging for income and the proportion of the income in the total fell to 17.5% in 2007. Restrictions of payment schemes had significant negative impacts on the income of farmer households and largely reduced the timber logging on the forestland of the villages.

The payment is the only compensation of the government for local farmers to stop timber cutting. The standard of the payment was 52.5 Yuan per ha from 2001 to 2005 and increased to 67.5 Yuan per ha in nominal terms in 2006 due to the establishment of the regional payment schemes. The amount of the payment which the farmer households received depended on the area of their public benefit forest. The average forest land of a farmer household in the villages was 8.27 hectares and the corresponding average payment for a household was 483 Yuan in 2007. However, the payment accounted for 4.9% of the total household income, but only equaled to 7.4% of the loss of timber income between 2001 and 2007. It was too low to compensate the loss of farmer households. This is line with perceptions of the farmer households. According to the farmer household survey, most of the farmers thought that the payment schemes had negatively affected their income; only two farmers believed their income had increased from forest related

*Figure 5.5. Bamboo forest plantation and logging: an important income source for local farmers in Guangxi.*

tourism after the schemes. It further confirmed the negative impact of the payment schemes on farmer household income.

Apart from the direct income loss, the forest use restrictions of the payment schemes also had negative impacts on fuel consumption and other agricultural production, such as reduced supply of fuel and prohibition of livestock browsing on forestland. Over 68% of the respondents thought that the schemes reduced fuel wood consumption and wood is the most important heating energy of rural areas in Guangxi's humid and cold winter. 21% indicated that their family animal husbandry was affected by the schemes; and also other difficulties in their life due to the schemes were mentioned (see Figure 5.6).

Through the analysis above, sizeable difference existed between the income loss and the additional payments. When the expectation of farmers was taken into account, the gap became even huge. The survey showed that farmers appealed to an average payment of 1,875 Yuan per ha, ranging from 300 to 4,500 Yuan per ha annually depending on the quality of forests and the value of forestland if the government still wanted to keep current size of public benefit forests. 76.5% of the farmers agreed on developing a biding scheme to decide the area of public benefit forest sand the payment on a voluntary base. 26% of farmers were willing to turn all of their own forests into public benefit forests if the government can provide sufficient payment. However, 21% of the farmers preferred to reduce the proportion of their public benefit forest to less than 50% of their forest.

### 5.7.3 Participation mechanism

This part discusses how farmers participated in policy formulation and implementation of payment schemes, concentrating on four stages: policy formulation, demarcation, management and examination of payment schemes.

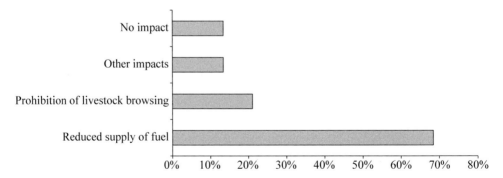

*Figure 5.6. Payment schemes' impact on fuel consumption and other agricultural production of local farmer household (n=15).*

*Payment policy formulation*

There was no formal direct channel for farmers and village committees to participate in formulating payment schemes. They often expressed their opinion on the payment schemes to township forestry stations, county forestry bureaus and local officials. Local forestry authorities represented local farmers by putting forwards their opinion to the government for discussion. Upon several issues, like raising payment standards and increasing public benefit forests, local forestry authority often aligned itself with local farmers, based on parallel interests. Besides the channel for opinion delivery through the administrative system, the delegates of the Autonomous Regional People's Congress (ARPC), who are elected by a multi-tiered representative electoral system, can make formal proposals for payment schemes during the meetings each year. However, this political process cannot ensure the proposals being put on the governmental agenda. The acceptance on the proposals depends on various factors such as financial feasibility, power of related authority, focus of the central government, and public attention. Since 2003, establishing a regional payment mechanism for public benefit forest has been proposed on the meetings of the Autonomous Regional People's Congress (ARPC) continuously. In 2006, the regional government set up a regional payment fund for the public benefit forest at the regional level. However, the fund was still too little to meet the needs of local farmers to manage their forest for ecological purpose. In 2009, the annual working report of the regional government, which had to be audited by the Autonomous Regional People's Congress (ARPC), started to mention exploring new payment schemes for forest ecosystem services but there was still no specified plan on its political agenda. Therefore, the suggestion for establishing new payment schemes was still brought forward to the meeting of the Congress every year and requested an official reply from the forestry department. In 2010, the forestry department received 14 proposals from delegates and 8 of them were directly related to forest ecosystem restoration and public benefit forest. Especially, 4 proposals asked to set up a payment scheme at the sub-regional or watershed level to complement the existing central and regional schemes and 3 proposals suggested to increase payment standards of current payment schemes for local farmers and state-owned forest farms.

*Demarcation of public benefit forest*

According to the "Measures on Demarcation of Key Public Benefit Forest" issued by the State Forestry Administration and the Ministry of Finance, consultation with local farmers is required in the process of demarcation. However, in practice, the regional government demarcated public benefit forest only based on technical standards and did not take the consultation of local communities and farmers into consideration. Officials in the regional forestry department explained that the tight time schedule resulted in reluctance of governments to allow participation of local forest owners and they also had expected increase of payment from the central government in the long run, which can benefit farmers and reduce their opposition in the end.

Village heads in our sample thought that the area of public benefit forests should be decided though a village meeting involving all members. The farmer household survey showed that 26% of the farmers thought that the demarcation should be decided by the governments (taking local situation into consideration or demarcating equal area of public benefit forests for each household),

37% of the farmers preferred the village committees and 37% preferred a village meeting. This response demonstrates that although local farmers still to some degree trusted that governments can make a fair and reasonable arrangement for demarcation, more of them hoped that the power of decision making could be transferred to communities or themselves.

The interviews with officials from regional forestry departments showed that in 2001 each county exclusively made a draft arrangement on demarcating public benefit forest by using national technical standards. Then, each municipality summarized the arrangements and submitted it to the regional forestry department. With the technical support of Guangxi Forest Inventory and Planning Institute, the regional forestry department validated the arrangement for demarcating public benefit forest. According to the interviews and group meetings, most local farmers have not been involved in the process of demarcation. Only 11% of farmer households were informed and consulted in a representative meeting. As a result, 84% of respondents strongly disagreed with the demarcation arrangement, and only 16% basically agreed to it. The major reasons for opposing the demarcation include too low compensation, failure in repaying contracts and fulfilling contracts, and reduced supply of fuel wood (Figure 5.7). For the interest of ecological restoration, only 24% of the farmers considered the demarcation arrangement reasonable. However, local farmers have a strong tendency to combine ecological interest with their income and livelihood and they would not consider ecological benefit alone. For example, plantations along the riverside are usually both profitable due to good soil quality and convenience for transportation, and ecologically important for preventing soil erosion. Local farmers generally paid more attention to economic benefits of such plantations and thought that it was unreasonable to include these into public benefit forest. The loss of timber income and low compensation ranked first among the reasons for disagreement. Getting into debt and breach of contracts was the second important reason, since many farmers have taken a loan from banks for establishing plantations and contracted forestland with villages. In addition, disagreement also focused on the negative impact on daily life such as timber need for household construction and fuel.

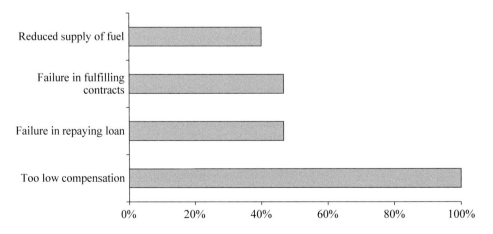

*Figure 5.7. Reasons for opposing the demarcation of public benefit forests (n=15).*

In Guangxi, local farmers have not been included in designing, modifying or fine tuning the demarcation arrangement and the low coverage of informing and consultation with local farmers also contributed to a high rate of opposition.

## Management of public benefit forests

Management of public benefit forests includes signing management contracts, distributing forest protection tasks and conducting daily forest management. The responsibility of management was distributed by two types of contracts to reach the local level. First the township government signed an administrative responsibility contract with village committees, under the terms of which village committees are in charge of distributing and monitoring public benefit forest protection tasks. The survey showed that 33% of farmers had been informed in the signing of responsibility contracts.

All forestland in the sample villages has been contracted out to farmer households since the first round of forest tenure reform in 1982. Therefore, the villages made a contract with farmer households for protecting their public benefit forest without appointing foresters. However, contracts for public benefit forest protection have not been carried out strictly and 61% of the farmers did not sign a formal contract with the village committees. The survey showed that only 39% of the farmers actively conducted daily management on their public benefit forest, such as mountain ranging, and most of the farmers just passively stopped timber logging. Most farmers (60%) who carried out daily management found the area of public benefit forest under their management too large, while 40% thought it was reasonable for them to manage. Furthermore, all the farmers considered payment too low, even only for forest management.

## Examination on the performance of the payment schemes

Although there was no formal policy for public benefit forest management at regional level, county forestry departments have the responsibility of examining performance of protection and management of public benefit forest annually. The examination focused on illegal logging on public benefit forest. The autonomous regional forestry department also carried out reexamination on the implementation of public benefit forest protection from time to time.

The survey showed that few farmers had been consulted for opinions and suggestions when forestry stations and high level governments conducted the (re)examination of public benefit forest management. However, 65% of the farmers often actively gave opinions or revealed problems in payment scheme implementation to village committees and forestry departments. The village committees were the first choice for the local farmers to deliver opinions and township forestry stations and the county forestry bureau were taken as the second and third important channels. Referring to efficiency of examination, local farmers thought that self-monitoring is the most effective and examination by the autonomous regional forestry department, county forestry bureaus or township forestry stations is less effective (Figure 5.8).

The local farmers had a strong tendency to manage and monitor the public benefit forest by themselves and at the same time lacked trust in forestry bureaus and township forestry stations. This distrust positively related with the low levels of consultation to local farmers during government-dominated demarcation and examination processes.

## 5.8 Discussion and conclusion

The case of Guangxi shows the relationship between the institutional setting of forestry management and protection and the payment schemes. Both state-owned forest farms and villages were selected as research objects. Forest ownership, organization structure and administrative status constructed the main institutional variables for payment schemes. They shape forest management into different styles in state farms and villages. State-owned forest farms rely on administrative management of public benefit forests, have more discretionary power on the use of their forestland and take payment schemes as a kind of subsidy for their management tasks. However, the management of public benefit forests at the village level involves various stakeholders such as governments, village committees and local farmers. Under the payment schemes, villages play roles in consulting on public benefit forest management and take part of the responsibility for its management. The payment from the schemes are regarded or expected by local farmers as compensation to the loss of local forest owners. Second, state-owned forest farm reform and collective forest tenure reform are imposing respective influence on the implementation of the payment schemes in state farms and villages. The reform on state-owned farms strengthens the centralized management on public benefit forests. The forest tenure reform however facilitates a decentralized management on public benefit forests. Third, classification-based forest management and public benefit forest management policies are major pillars, which serve to politically secure the public benefit forest and make a division between commercial and public benefit forests. They also stipulate different responsibilities for state-owned farms and villages in the management of public benefit forest. Finally, forest industrial policy influences the payment schemes in both positive and negative ways. It transfers the timber demand originally resting on public benefit forests to commercial forests at the provincial level, but it also brings negative incentive for public benefit forest owners by enlarging the gap of economic returns from public benefit forests and commercial forests.

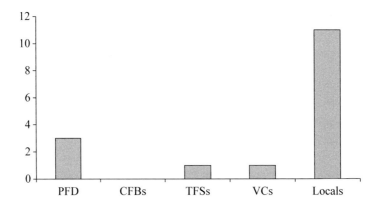

*Figure 5.8. Frequency of preference to public benefit forest examination body (n=15).*
*PFD: provincial forestry department; CFBs: county forestry bureaus; TFSs: township forestry stations; VCs: village committees; Locals: local farmers.*

The payment schemes have been implemented in Guangxi for more than ten years. They also in turn have influenced the institutional setting. Overall, ecological rationality is being built into the institutional setting through the implementation of the payment schemes. From the reform on state-owned farms and the collective forest tenure reform, to forest industrial policy making, ecological conservation is taken into account and even given priority in policymaking. During their interaction, payment schemes and classification-based forest management facilitate the reform on state-owned forest farms, which specializes the organization structure of some state-owned forest farms for the purpose of ecological conservation. In villages, payment schemes push for the tenure reform on public benefit forest.

The payment schemes fundamentally changed forest use practices of state-owned forest farms and villages. The case of the 3 forest farms shows that state-owned farms already partly shifted from logging forests for timber production to implementing ecological conservation in their forest ecosystems. Two of them are changing their organizational structure from timber production enterprises to public benefit forest management agencies. The investigation in the villages demonstrated similar trends in the forest management. Forests with important ecological value have been protected from logging. Some contextual factors also differentiate this transformation of forest use practices among different state farms and villages. According to the evaluation, two factors stand out on account of their influences. First, the natural geographical condition provides different opportunities for state farms or villages. While some state farms can take advantage of using their unique forest ecosystem to develop ecotourism and easily transform their forest utilization from timber logging to ecological conservation, other state farms have no potential to develop ecotourism and suffer more from economic loss caused by the payment schemes. Second, dependence of local livelihood on forest resources also plays an important role in the shift of forest use practices. Less dependence on forest resources and more opportunities for off-farm work and other income sources facilitate the change of forest utilization in villages.

According to the evaluation in Guangxi, the payment schemes have positive effects on environmental services. However, the severe snow disaster struck the public benefit forest in 2008 and reduced its ecosystem services to a large extent. Comparatively, state-owned forest farms made more active efforts in managing and protecting public benefit forests than villages. Payment schemes did not succeeded in providing sufficient compensation for both forest farms and local farmers. The state-owned forest farms are struggling to maintain the management on public benefit forest by "borrowing" from their other revenues. The income of local farmer households has been negatively impacted by the payment schemes, especially the timber income. In addition, the standard for payment should be diversified according to different environmental values and costs related to the forests. For state owned forest, a cost-benefit analysis can be employed to provide more accurate information for the payment. For the forests under management of many farmer households, a bidding scheme for payment could be a more efficient way to decide the payment for public benefit forest.

The state-owned forest farms have been apparently involved in the policy formulation, demarcation, management and examination on the public benefit. However, the local farmers had limited channels to participate in the implementation of the payment schemes. A more open participation mechanism, which can better inform and consult local communities and farmers, should be set up to reduce local resistance and improve the legitimacy of the payment schemes.

The assessment on payment schemes in Guangxi demonstrated the difficulties in the implementation of payment schemes in least developed regions. Local governments did not have sufficient financial support for developing new local payment schemes and formal regulations on public benefit forest management have not been in place. Severe negative impacts on the local economy and farmers' income increased the complaints and rejection from local farmers. Low administrative capacity and efficiency, which is relatively universal among the governments of least developed regions, cannot provide enough support for monitoring and implementation of the payment schemes.

The implementation of payment schemes for forest ecological services in Guangxi improved its forest protection, in the sense that timber logging has been largely limited in public benefit forest, ecological priority has been built in its forest policy making and economic incentive works with regulatory instruments to facilitate ecological conservation in the forest sector. Forest use practices of both state farms and local villages are changed by the payment schemes. The change can be interpreted as a process of ecological modernization in Guangxi's forest sector. The forest sector has been restructured from timber production plan management to classification-based management which puts emphasis on both commercial and ecological purposes. Ecological rationality has been built into the process of forestry production and management: public benefit forest management has been integrated as the major task of forest departments at all levels; and local forest use practices have been changed to serve ecological objectives. However, this transition is driven mostly by administrative responsibility and top-down implementing methods. Participation is not in place to guarantee acceptance of local people on the payment schemes. Some local communities and farmers are further marginalized through the implementation of payment schemes. In the long term, such lack of consensus between governments and local communities may put payment schemes and the PBF at stake.

# Chapter 6.
# Payment schemes in the transition of collective forest tenure in Liaoning[32]

## 6.1 Introduction

Northeast China is a region abundant with forest resources and was regarded as one of the important timber production bases in the country. Among the northeastern provinces, Liaoning Province was the first one to be included into the central payment schemes and a pioneer to carry out forest tenure reform. Its forest sector experienced drastic change along a process of ecological modernization, in which ecological rationality became increasingly built into forest management. Payments schemes for forest ecological services play an important role in reforming its forest management system. Especially, compared with other northern provinces, Liaoning Province depends less on funding from the six national key forest projects and puts more efforts to develop its own provincial payment schemes. The development demonstrates a colorful vision for the implementation of payment schemes in China northern regions. In this chapter, I first give a brief introduction on the natural conditions of Liaoning Province. In Section 6.2, the emphasis is put on explaining the institutional setting for payment schemes in Liaoning Province. Section 6.3 introduces current payment schemes implemented in Liaoning Province. After that, the performance of payment scheme is evaluated. This analysis is based on a case study carried out in two sample counties. The last section presents the results of the evaluation and discusses how to interpret the implementation of payment schemes in Liaoning Province.

Liaoning Province, in the southern part of China's northeast, consists of 14 municipalities and 74 counties and districts (Figure 6.1). With 4,641 thousand hectare its forest coverage is 31.84%. However, the forest resources are unevenly distributed over the province. Over 69.1% of its forest area, 82.6% of the forest stocking volume, and 90% of its natural forest concentrates in the eastern part and most natural forest functions, such as watershed conservation, are also to be found in this region.

Under a continental monsoon climate, Liaoning Province is fragile following frequent droughts and floods. Deforestation for crop planting and development further deteriorates the natural environment and leads to more frequent natural disaster incidents in the region (Zhang *et al.*, 2005). Besides uneven precipitation and a mountainous terrain prone to erosion, anthropogenic factors worsened soil erosion, including farming on sloping land, logging for firewood, overstocking wild silkworms (tussah), and mining (Wang, 2008). The frequency of floods in the eastern part of Liaoning Province has changed from once every 10 years during the 1950s to once every 3 years after the 1980s. Droughts also more often struck the region since 1980. Furthermore, soil erosion extended over the province, covering 5.1 million hectare and even in the more forested

---

[32] This chapter is based on an article submitted to Journal of Environmental Policy and Planning in May, 2011, as Dan Liang and Arthur P Mol, Political modernization in China's forest governance: Payment schemes for forest ecological services in Liaoning (under review).

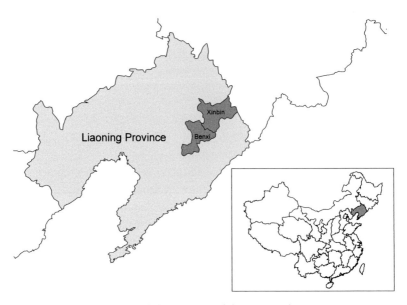

*Figure 6.1. Liaoning Province in China and the location of the case study counties.*

eastern part, 30% of the area has been impacted by soil erosion. The large and medium sized water reservoirs in Liaoning lose around 100 million cubic meters of storage each year due to severe soil erosion. In addition, severe desertification still endangers ranches and farmlands in the western part of the province.

Facing serious environmental degradation, the central government established a pilot funding mechanism to encourage local governments (including Liaoning Province) to sustain and recover forest ecosystem services from 2001 to 2004. At the same time, the central government promoted a leapfrogging development in forestry, moving its focus from timber production to ecological conservation. Six national key forest projects have been introduced to fulfill this turn (see Chapter 3). The basic logic of these projects is to increase forest cover by large scale afforestation and conserve forest ecosystem by large national investments. The Central Fiscal Forest Ecological Benefit Compensation Fund Program (FEBCFP) also served for this turn in China's forestry development. In 2004, the fund was formally set up nationwide, which built "paying for ecosystem services" into China's forest management and protection. Along with the national payment scheme, some local payment schemes have been formulated to cover more forest and help to secure its eco-services. After they have been implemented for more than 6 years, the local payment schemes show a more stable patterns than the Six Great Projects, which are often criticized for absence of institutionalization (Wang *et al.*, 2010). Once these national projects end as planned (NFPP in 2010 and CCFGP in 2021), the natural forest will again face threats from logging and conversion to cropland.

## 6.2 Institutional setting for payment schemes in Liaoning province

Although severe environmental problems relating to deforestation emerged in Liaoning Province, the province did not initiate action to deal with them. The central government actually played the role of promoting forest protection and conservation among the provinces and urged them to separate public benefit forest from commercial forest, to formulate regional regulations on forest management, and to share the responsibility of funding payments. China's decentralization following the marketization reform in 1979 and the recentralization during the 1990s continuously reshaped the central-local relations. Central-local relations are characterized as concentrated fiscal resource at the central government and discretionary power of local authority in decision making, planning and implementation. These central-local relations formed a basic institutional setting for China's public benefit forest management and emerging payment schemes.

### 6.2.1 Institutional arrangement for public benefit forest management

In 1984, China's first forest law started to adopt forest management strategies based on a classification system of forest resources in order to combat deforestation and degradation of forest ecological functions. The classification system includes five categories of forests: protection forest, timber forest, economic forest, fuel forest and special use forest. It aims at specifying the use of forest resources, applying different management measures to different forests, and further reducing overcutting. The division between different categories is controversial, since any forest might provide multiple services. However, it indeed created the possibility to restrict the discretion of local forest authority on forest use, which usually preferred logging forest for timber. In addition, the classification system helped ecological functions to stand out as being at least as equally important as other functions of forests for local officials, forest managers and the public. As such it has been widely used in forest resource survey, statistics, and planning. But the majority of forests in China were still classified as timber forest (Dai *et al.*, 2009). Not surprisingly, when the central government only makes principles and rules but the local government exercises discretion on forest planning and classification, local governments tend to classify forest mainly as timber forest.

In 1988, China's forest law was amended to reclassify the five forest categories into two: commodity forest and public benefit forest for ecological purpose (Dai *et al.*, 2009). This two-category forest classification system juxtaposed ecological functions with economical commodity purposes to further emphasize the importance of ecological attributes of forest. This reclassification also created convenience for forest practitioners when operating in local forests. Their focus only need to concentrate on whether a plot of forest should be protected for ecological functions or utilized for commercial purpose. However, the incentive for local forest authority to enlarge the public benefit forest area for ecological purpose was still not in place. Moreover, after the 1994 fiscal reform, the central government established its own tax collection system and seized a majority share in the tax base. Therefore, local governments suffering from a shrinking tax base have no sufficient resources and capacity to support larger areas of public benefit forests. However, a set of technical criteria for forest classification and management measurement has been formulated and tested in 10 forestry bureaus selected by the central government in 1996 (Dai *et al.*, 2009). From 1996 to 2000, the forest administration of the central government struggled to channel funding for extending public benefit forest nationwide by negotiating and bargaining with other ministries. In 2001, 1 billion Yuan has been mobilized from the Ministry of Finance to set up a pilot payment scheme for forest ecosystem services in 11 provinces, including Liaoning.

Since then, the central government has issued a series of regulations to direct the management of public benefit forest, such as *Measures on Demarcation of Key Public Benefit Forest* (State Forestry Administration and Ministry of Finance, 2004) and *Management Measures for the Central Compensation Fund for Forest Ecological Benefit* (Ministry of Finance and State Forestry Administration, 2004). Following these national rules, Liaoning Province set up its own management institutions for public benefit forests – *Liaoning Province's Detailed Rules for Implementing the Management of Public Ecological Benefit Forest* (brief for Detailed Rules) in 2006, which stipulates formal rules for classification, protection, productive management, monitoring and inspection, and punishment for noncompliance (Liaoning Provincial Forestry Department, 2006a).

In China, agreements between governments serve to link requirements in the central plans, programs and projects to local governmental policies and officials, but also to define and implement goals and requirements between and within governments and agencies (Guttman and Song, 2007). The target responsibility system, which includes these agreements or contracts, operates as an important component of the payment schemes. In general, the target responsibility system used by governments involves a set of targets, a point system, responsibility certificates and provisions for monitoring of performance (Guttman and Song, 2007). Although it varies from province to province, the target responsibility system for public benefit forest management usually includes multiple targets such as institutional building, public benefit forest protection, the use of payment, and information system management. Each of these targets is represented by one or several quantitative and/or qualitative indicators. For example, whether or not a special organization is established to take charge of public benefit forest management is one of indicators for institution building; and the rate of forest loss to fire partly refers to the target of public benefit forest protection. Criteria and corresponding points are also employed to measure the targets. For instance, for the rate of forest loss to fire, the points will be deducted if it is more than 1‰. The responsibility of public benefit forest management and protection is distributed within the multilayered state structure (the central level, province, city, county, and township) by the agreements along each level. The leaders of forest departments at each level are de facto responsible for public benefit forest management and protection, although the leaders of governments generally should take the main responsibility. Public benefit forest management and protection is only one of the many targets that provincial and local officials must fulfill and it is not as important as other economic targets such as GDP requirements. A point system weights the importance of each target. As a result, limited resources can be mobilized from governments for public benefit forest management and protection. Few county forestry bureaus established specialized divisions for public benefit forest management and these tasks are generally separated and distributed over different divisions as part time jobs.

### 6.2.2 Collective forest tenure reform in Liaoning Province

Liaoning Province started its collective forest tenure reform in March 2005. It is one process of decentralization, in which collective forestland is distributed to individual farmer households, delimited with clear boundaries and under protection of legally-binding certificates. The objective of the reform is to turn forest resources into production factors which can be exchanged and deployed in a market-based economy and to further improve the efficiency of forestry production.

It is controversial among forestry authorities to include public benefit forests into the tenure reform (State Forestry Administration, 2006a). The central government did not provide uniform guidelines for the implementation of tenure reform on public benefit forest and allows each province to decide whether the reform covers public benefit forest. Some local officials (in Fujian and Guangxi Province) argued that it would be hard to monitor the management of public benefit forests if it is distributed to individual farmer households. In the past, the responsibility of managing and protecting public benefit forest resided with villages. The forestry bureaus only needed to make a responsibility contract with each village. But they have to directly face with hundreds and thousands of farmer households if the public benefit forests are handed out. It is often questioned whether local forestry bureaus have the capacity to monitor such numerous small forest plots. Furthermore, the official documents from the central government to local governments emphasize that improving economic efficiency of forestry production and increasing income of local farmers justifies the tenure reform. However, under current low payment standards, public benefit forests are often regarded as a burden on owners, rather than an income source. Therefore, inclusion of public benefit forests into the reform has not added any incentive to farmer households to better manage and protect public benefit forests.

However, the proponents for extending the reform to public benefit forests think that the reform is a good opportunity to settle down the responsibility of managing public benefit forests. After commercial forests are redistributed to farmer households, public benefit forests will become a common pool susceptible to illegal logging for timber or firewood. This side effect of the reform has already happened in some southern provinces, such as Fujian Province which excluded public benefit forests from the tenure reform. Local officials in Fujian and Guangxi (based on the interviews in Fujian and Guangxi) often complained that the redistribution of commercial forests had imposed pressure on public benefit forests and increased the difficulty of protection and management. In understanding the risk behind the reform, Liaoning Provincial Forestry Department experimented with a comprehensive arrangement to extending the tenure reform to public benefit forests and tried to share the responsibility with local farmers. In the past, local farmers were often regarded as threats to the public benefit forest and the governments tried hard to separate them from the public benefit forest and constantly monitored them with local forest stations and villages. Therefore, the decentralization of the ownership of public benefit forests is a process of facilitating participation among farmers, who start to be treated more as cooperative partners for protection than as potential threat to forest. At the same time, responsibility, obligation, and rights are also redefined and redistributed among governments, villages, and farmers (Figure 6.2). Before the tenure reform, the public benefit forest was managed by village committees. Therefore, forestry bureaus made direct responsibility contracts with village committees for monitoring public benefit forest management. At the same time, forestry bureaus hired foresters to protect the public benefit forest from fire, pest, disease and illegal logging. Local farmers had no access to the public benefit forest. After the reform, the forest tenure rights were transferred from village committees to local farmer households. Individual farmers gained the rights to use their family plots of the public benefit forest to develop agro-forestry and other production as long as it did not hinder ecological service of the public benefit forest. They also have to share the responsibility of public benefit forest protection through a management contract with village

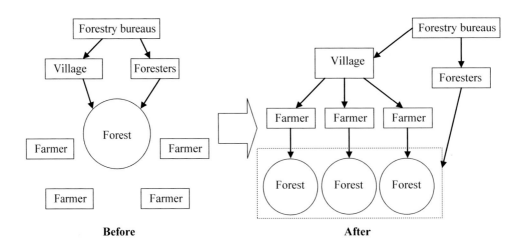

*Figure 6.2. Change in forest management after the tenure reform.*

committees. Depending on their willingness and the size of their family plots[33], they can decide to delegate the task of public benefit forest protection to foresters or perform it by themselves. Through the tenure reform, local farmers moved from a relatively marginal position into the center of public benefit forest use and management.

The second reason justifying redistribution of public benefit forest is the opportunity to develop agro-forestry in the public benefit forest ("Lin Xia Jing Ji" in Chinese). Early in the 1950s, China started to implement multiple-use forest management, but it failed in supplying sufficient timber products and ecological services to society. The superficial emphasis on multiple use of forest turned into de facto concentration on timber harvesting in practice. After half a century of unprecedented efforts driven by both timber and environmental demands, China is still able to produce only 50-80 million m³ of industrial wood annually[34] and at the same time, the forestland and its associated environment have been deteriorated (Zhang, 2005). The classification-based forest management was an overall counter attack to previous multiple-use forest management. After the devastating flood in 1998, many provinces adopted a strict protection measure on forest including completely banning timber harvesting, prohibiting grazing in forest, and even restricting access to forested region. For instance, a lot of bridges and roads to forested region in Sichuan Province were demolished to avoid disturbance from human activity. Following a series of forestry ecological restoration projects from 1998 to 2001 such as the Natural Forest Protection Program (NFPP), the Wildlife Conservation and Nature Reserve Development Program (WCNRDP) and payment schemes supported by central and local governments, this classification-based forest management has been applied nationwide. This forest management paradigm has advantages, such

---

[33] Usually, if forest plots are too small and fragmented, farmers will delegate the task of forest protection to foresters.

[34] With similar areas of commercial forest, the United States of America produces about 480 million m³ of industrial wood, more than 6 times China's production (FAO, 2010).

as easy to implement with low requirements for management capacity. However, its dichotomy of forest resource also has been criticized as it neglects multiple coexisting functions of the forest ecosystem and it lowers efficiency of the forest ecosystem for other production such as non-timber forestry products and agro-forestry. Furthermore, the implementation of classification-based forest management deprived local communities and farmers of forest resources, which they have been dependent on for generations. Local communities' need for sustenance and development induced debates on the classification-based management paradigm. Is forest ecosystem conservation essentially incompatible with local livelihood and development under current rural socio-economic situation? Can they be combined together and improve each other under some management paradigm? The interviews with local officials and village members in Liaoning Province showed that under the condition of sustaining a healthy forest ecosystem, it was possible to develop agro-forestry in the forest such as traditional Chinese herb medicine planting, fruit and nuts harvesting, animal grazing and tourism, and to improve the livelihood of local farmer households. This new management needs to take into account the special natural conditions in each forested region and a lot of efforts should be input to seek suitable development opportunities, to supervise the use and management of forest, and to govern the degree of development in public benefit forest. Some counties such as Benxi County have formulated related provisions in the public benefit forest management measure to govern and guide non-timber use of public benefit forests and avoid overusing and ecological degradation (Benxi County Government, 2005).

Thirdly, the reform is a process of both marketization and the rule of law. The idea of a market-based economy, forest ownership and forest as capital assets is accepted by local farmers. Clarification of property rights on forest provides an institutional space for marketizing environmental services from forest ecosystems. The government also established a financial supporting system to facilitate the marketization of forest resources. In May 2009, Shenyang Branch of China People Bank and Liaoning Provincial Forestry Department co-issued a notice *Opinion on Promoting Loans Secured by Forest Tenure*, which stipulated that national and regional public benefit forest can be used as collateral for a secured loan from banks. Until January 2010, the total loan secured by forest tenure has reached over 1.6 billion Yuan (Liaoning Provincial Forestry Department, 2010). However, according to officials from the forestry department, in practice it is much harder for the owners of public benefit forest to apply for secured loans from banks than those having commercial forest, although the notice accepted both types of forest as collateral. Compared to commercial forest, it is difficult to evaluate the value of public benefit forest and to resale it for paying back the debt if the borrower defaults. Although the Forest Law and the Rural Land Contract Law allowed forest resources to circulate by market, detailed regulation is still not in place and current regulations and policies on forest tenure cannot provide sufficient support for market-based circulation of public benefit forest. In 2010, Benxi Municipal Government issued *the Regulation on Forest Circulation* to govern the procedure of forest tenure circulation including public benefit forests. It is still in an initial stage for the marketization of environmental service of public benefit forests. The sketch already showed the pivot of market in attracting cash flow into management system of public benefit forests for environmental services. In addition, the forest tenure reform was also an important force to push the central government to raise the standard of the payment from 75 to 150 Yuan per ha in 2009. The reform enlarged the gap between the profit of commercial forest and the payment for public benefit forest. It aroused dissatisfaction from local farmers for the payment schemes. Clarifying forest tenure also improved the rule of law in forest sector. The awareness of farmers on forest property rights was enhanced. They strongly requested the central government to raise the payment standard to compensate their loss due to the logging ban, or to change their public benefit forest into commercial forest. Local officials, who have more channels to express opinions to governments, also suggested that the management and protection of public benefit forests became more difficult since the collective tenure reform and a big rise in the payment was required to appease local farmers. Therefore, the forest tenure reform not only created an institutional setting for marketizing environmental service of public benefit forests, but also improved the acceptance of the whole society, including governments and farmers, on property rights of public benefit forests, that provides the environmental service.

### 6.2.3 Forestry industrial development as a solution

Facing local economic downturn and negative livelihood impacts resulting from a logging ban on public benefit forests and natural forests, the provincial government expected forestry development to revitalize the local economy and improve the livelihood of farmers under the condition of sustaining a healthy forest ecosystem (Liaoning Provincial CCP Committee and Liaoning Provincial Government, 2009). Before 2003, the issues on the negative impact of public benefit forest protection on local economy were brought forward to the "Two Meetings" (the People's Congress and the People's Political Consultation Conference) by the representatives at the provincial level. In 2004, the Provincial Forestry Department (PFD), the Policy Research Office of the Provincial CCP Committee, and the Policy Research Office of the provincial government, together carried out a policy investigation in the forested region, which had been severely impacted by the logging ban. A series of reports were generated and delivered to the Provincial CCP Committee and the provincial government. Reacting on the reports, the PFD provided a principle "industrializing ecological conservation and ecologizing industrial development". Actually, it is in line with ecological modernization and articulates that ecological rationality should be built in forestry production. This principle coupled forestry development and ecological conservation, and proposed a solution to the conflict between local economic development and forest ecosystem protection. A consensus was reached between different departments of the provincial level, mainly including the Financial Department, the Development and Reform Commission, and the Forestry Department. Following a top-down way, local governments, which were already in a severe financial shortage and often had a strong preference to economic development, more than welcomed such an idea. A lot of examples about agro-forestry were picked up by local governments and disseminated by the media with the support of the provincial government. This was a communicative process in which the initiatives of villagers and companies were taken as exemplars to materialize the principle "industrializing ecological conservation and ecologizing industrial development". Then, the government summarized different models of agro-forestry from the successful examples and promoted them to wider rural areas depending on local context. There are three types of models: traditional Chinese medicine planting in forests; vegetable cultivation in forests; and livestock/wild animal breeding in forests.

Simultaneously, the government made plans to direct the development of agro-forestry and established supporting policies for it. In 2004, the provincial government issued the planning outline of ecological construction and the planning outline of forestry industrial development. Whereas the first outline stipulated different ecological requirements and ecological conservation strategies for different regions, the second indicated forestry development planning for different regions and the related policy orientation. From 2005 to 2006, the provincial government issued a notice about *speeding up forestry development in the eastern region of Liaoning Province*, which included specified supporting policy measures such as funding, subsidy, and preferential loans for forestry development projects (Liaoning Provincial Government, 2005b). It has to be noted that the support was only available for medium or large scale forestry development projects, not for small scale farmer household economies. Since 2005, the provincial government provided more than 15 million Yuan annually for supporting agro-forestry development. In 2008, under the support of the provincial government, the provincial forestry department renewed its forestry development plan–*Liaoning Province Development Plan for Doubling Forestry and Related Industry* (Liaoning Provincial Government, 2008). In that year, the subsidy from the government reached 50 million Yuan. These plans served not only as overarching rules for developing agro-forestry especially in public benefit forests, but also channeled funding resources for its development.

## 6.3 Payment schemes in Liaoning Province

The payment schemes define a field of public benefit forest, which is separated from the commercial forest and is supported by public subsidy. In Liaoning Province, this field includes three types of public benefit forests: national public benefit forest, provincial public benefit forest and provincial natural forest. The definition of public benefit forest is determined by forestry departments at the provincial and country levels. Under the administration of forestry departments, the distinction between a public benefit forest and a commercial forest is relatively clear and the public benefit forest is managed independently without overlap by other programs.

The payment schemes stipulate a series of technical functions and basic rules of procedure concerning public benefit forest protection. In Liaoning Province, the public benefit forest mainly functions as water source conservation, soil erosion prevention, wind shelter and nature reserve. Under the principle of ecological priority, Liaoning Province permits developing non-timber forestry production in public benefit forests including seed collection and agro-forestry. It gives room for economic utilization of public benefit forests and ameliorates the conflicts between economic and ecological functions. Management tasks and responsibilities are divided among county governments, forestry bureaus, township forest stations, village committees and forest rangers. Some basic rules of procedure are also set by the payment schemes, including public forest management organizations, monitoring and examination of forest rangers, forest resource management, funding management, and incompliance and punishment. Four principles are manifested in Liaoning Province's payment schemes: ecological priority, economic incentive, scientific utilization and administrative responsibility. The application of ecological priority is similar to other provinces. Any production in public benefit forests cannot harm ecological functions of the forests. Each county stipulates the detailed rules of procedure for different utilization of public benefit forests.

The principle of economic incentive is an innovation in China's forest protection, which facilitates public benefit forest management in rural areas usually stricken by poverty. Liaoning

Province already initiated three types of payment schemes to support protecting forest ecosystems (national payment scheme, natural forest protection scheme and provincial payment scheme). In 2001, the central government enlisted Liaoning as a pilot province for the central fiscal payment scheme for forest ecological services with an annual funding of 105 million RMB Yuan for 1.4 million hectare of forest. In the same year, the province launched a project to ban timber logging on natural forest for commercial purpose in 9 prime forested counties. This project, which adopted the same measures as the national Natural Forest Protection Project (NFPP), covered 787 thousand hectare of natural forest with 41.3 million Yuan for forest management and protection and 56.86 million Yuan transferred from the provincial finance to the county finance for compensation every year. In 2004, following the formation of the national payment scheme, the province established its own provincial payment scheme and earmarked 18.27 million Yuan for protecting 406 thousand hectare of forest. In total, 2.65 million hectare of public benefit forest and natural forest has been included into the payment schemes, accounting for 41.8% of its forestland.

The payment schemes in Liaoning also emphasize the principle of scientific utilization of public benefit forests. Forest owners are allowed to develop agro-forestry in public benefit forests under the guidance of county forestry bureaus and township forest stations. Livestock can browse in the middle-aged public benefit forest during certain months. This principle, which is combined with ecological priority, creates room for local forest owners to utilize their forest resources without doing harm to forest ecological functions.

Similar to other provinces, the implementation of payment schemes in Liaoning also relies on administrative responsibility system, in which the responsibility of public benefit forest management is consigned from provincial to local forestry authorities. Liaoning Province uses a set of strict rules to guarantee the commitment of resources to public benefit forest management. Township forest stations employ forest rangers to manage and protect public benefit forests and county forestry bureaus provide regular training and examination for them. The payment is distributed and used through financial and forestry departments according to a set of treasury management measures.

According to the principles of payment schemes, the introduction of payment schemes into forest protection policies could be framed as a practice of ecological modernization, where governments gradually resorted to incentive-based measures instead of solely relying on coercive policy instruments, and where scientific utilization with ecological priority replaces a one-sided pursuit of economic benefit of forest resources.

## 6.4 Performance of the payment schemes

The establishment of payment schemes is an important milestone of Liaoning's practice in forest management and protection. A vast area of forest has been transformed from commercial purpose into ecological ends with the momentum gained from the payment schemes. The forest tenure reform, which is widely regarded as a decentralization process to entitle individual farmer households to collective forestland, had a profound influence on local forest management practice. Therefore, it becomes imperative to examine the outcome of the payment schemes and explore the impact of forest tenure reform on public benefit forest management.

### 6.4.1 Introduction to the case site

To understand and assess the implementation of payment schemes two typical case study areas in Liaoning province have been selected, taking into account the following criteria: the starting time of payment schemes (the earlier, the better), the area of the public benefit forest, forests with high ecological and economic importance, and existence of forest tenure reform. Benxi County and Xinbin County[35] proved to be excellent case study area, fulfilling best these criteria. Benxi and Xinbin County, respectively under administration of Benxi City and Fushun City, lie in the main forested region of Liaoning Province. The two counties are close to each other and both located along Taizi River, which is the primary branch of Liao River. Most public benefit forest of the counties is used for conserving water sources and preventing soil erosion (Table 6.1). Since 2001, these two counties have been included into the pilot of the central fiscal payment scheme. From 2005, collective forest tenure reform also has been implemented in the case counties. It was the earliest trial of tenure reform on public benefit forest nationwide. In addition, Benxi and Xinbin County have sizeable Chinese traditional medicine planting and mushroom and fungi cultivation. These industries have profound impacts on local forest resources and livelihood of farmer households.

Four villages were selected from the two counties using similar sampling criteria as for the county (Table 6.2). All four villages are located in mountainous areas but have good access (with cement roads) to towns and cities. Yanghugou Village ranks first in area, population size, forest resources and public benefit forests. Beiwangqing Village has a similar socio-economic profile as Yanghugou but has less forestland and public benefit forest. Dayang and Luoquan are quite similar in area, forestland and public benefit forest and both have relatively higher income per capita than the other villages (Figure 6.3).

*Table 6.1. Socio-economic situation of the two sample counties.[1]*

| County | Township | Area (km²) | Population (1000) | Forest coverage (%) | Public benefit forest (km²) | |
|---|---|---|---|---|---|---|
| | | | | | national | provincial[2] |
| Benxi | 14 | 3,344 | 300 | 73 | 1,203 | 943 |
| Xinbin | 15 | 4,287 | 310 | 73 | 1,153 | 1,234 |

[1] Source: county forestry bureaus.

[2] Provincial public benefit forest includes natural forest under the Liaoning Province Natural Forest Protection Program.

---

[35] The full names of these two counties are Benxi Manchu Autonomous County and Xinbin Manchu Autonomous County.

*Table 6.2. Socio-economic situation of sample villages.*[1]

| County | Village | Area (km²) | House-hold | Population | Crop-land (ha) | Income per capita (Yuan) | Forest land (ha) | Public benefit forest (ha) national | Public benefit forest (ha) provincial |
|---|---|---|---|---|---|---|---|---|---|
| Benxi | Yanghugou | 58 | 378 | 1,397 | 168 | 4,400 | 5,467 | 4,400 | 367 |
| | Dayang | 24 | 297 | 1,075 | 153 | 6,500 | 2,133 | 800 | 400 |
| Xinbin | Beiwangqing | 52 | 409 | 1,380 | 375 | 4,000 | 4,520 | 546 | 1,763 |
| | Luoquan | 28 | 168 | 621 | 38 | 5,000 | 2,580 | 433 | 700 |

[1] Source: interviews with local village leaders.

*Figure 6.3. Dayang Village, Benxi County.*

### 6.4.2 Implementation of survey

In-depth interviews and surveys have been employed to collect information and data. In understanding the general implementation of payments schemes and collective forest tenure reform in Liaoning, in-depth semi-structured interviews have been conducted with 4 officials from the Ecological Benefit Forest Management Office and the Forest Tenure Reform Office in Liaoning Provincial Forestry Department. Interviews were also carried out with 5 officials of municipality and county forestry bureaus to understand the implementation of payment schemes at the county level, to collect related documents and to select sample villages.

After the interviews, a survey of farmer households in 4 villages of Benxi and Xinbin County was conducted (Table 6.3). Between 12 and 15 farmer households have been randomly sampled in each village. A total of 54 farmer households were interviewed and 54 useful questionnaires were obtained (Figure 6.4). The farmer household survey asked farmers about their income and livelihood, forest resources, participation in payment schemes and forest tenure reform, attitudes toward the payment schemes and the reform, and willingness for compensation and adjustment. Two sample counties are minority regions and 83% of respondents are Manchu. All respondents completed primary education and 85% finished middle school education. 61% of interviewed families have 4-6 members and 89% of the households have less than two children (younger than 18 years old).

*Table 6.3. Characteristics of farmer respondents.*

| Residence | Yanghugou | Dayang | Beiwangqing | Luoquan | |
|---|---|---|---|---|---|
| | 13 | 15 | 12 | 14 | |
| Gender | male | | female | | |
| | 54 | | 0 | | |
| Ethnic group | Han | | Man | | |
| | 9 | | 45 | | |
| Age | 20-39 | 40-49 | 50-59 | ≥60 | |
| | 19 | 20 | 12 | 3 | |
| Education | Senior | | Junior | | primary |
| | 8 | | 38 | | 8 |
| Family size | ≤3 | | 4-6 | | ≥7 |
| | 20 | | 33 | | 1 |
| Children | ≤1 | | 2 | | ≥3 |
| | 48 | | 5 | | 1 |

*Figure 6.4. Interviews with local farmers at the meeting room of Village Committee of Beiwangqing, Benxi County.*

### 6.4.3 Payment policy evaluation

The evaluation for payment schemes includes three parts: environmental effectiveness, livelihood impact assessment and participation analysis.

*Environmental effectiveness*

According to the farmer household survey, 46% of the farmers believed their villages had no apparent soil erosion before the payment schemes started. About 33% of respondents thought their village had some soil erosion and 20% thought that widespread soil erosion existed before the schemes. In Dayang Village and Beiwangqing Village most of the farmers thought their village had some soil erosion. The respondents in Luoquan Village indicated no apparent soil erosion before the schemes, while about half of the respondents of Yanghugou Village thought that the village had widespread soil erosion. Annual precipitation in Yanghugou Village is over 800 mm, more than the other sample villages (740 mm) and 80% of the precipitation falls in July and August. In addition, steep sloping land extensively exists in Yanghugou Village. Therefore, it is the most fragile to soil erosion among the sample villages.

Around 24% of all respondents thought that soil erosion had been significantly reduced since the introduction of the schemes; 46% of them thought that some reduction had taken place and 30% saw no change in the local environment following payment schemes. When ecological improvement through the schemes is assessed, different baselines (degree of soil erosion before the schemes) should be taken into account. Almost all the respondents of Luoquan Village, where no apparent soil erosion had previously existed, thought that schemes brought no change to the local environment. In contrast, almost all the respondents of Yanghugou Village thought soil erosion has slightly been reduced, where widespread soil erosion existed under an ecologically fragile environment. In Dayang and Beiwangqing Village, most of the respondents thought soil erosion was reduced. Therefore, from the perspective of local farmers, the payment schemes achieved environmental effectiveness in Liaoning's rural areas in the sense that soil erosion was reduced in ecological fragile areas and ecological quality was maintained in areas less stricken by soil erosion.

Although there is no on site quantitative data available to show the change in public benefit forest resources after the payment schemes, the data at the provincial level shows that overall the public benefit forest in Liaoning Province increased stably and even faster in recent years (Liaoning Provincial Forestry Department, 2006b; 2009). Under the payment schemes, forest coverage in the public benefit forest increased from 77% in 2005 to 79% in 2008; the growth of stock accelerated with a rate of 5.4% in 2008 (Table 6.4). However, the extension of forest area and increase in stock volume caused a decline in the age structure of the public benefit forest in which young and middle aged trees dominate (92%) (Table 6.4). Provincial forestry officials indicated that this problem should be considered in policymaking of the payment schemes. A young age structure makes a forest vulnerable to the threat of extreme climate and forest pest and disease. In addition, the data from the farmer household survey shows that the forest tenure reform generated afforestation incentives for local farmers. According to the survey, 19% of the respondents have afforested on public benefit forestland, as the government provided a subsidy of 750 Yuan per ha for afforestation.

According to the governmental report, main threats to public benefit forest include forest fire, pests and disease, and illegal logging and occupation (Table 6.5). Whilst forest fire has been controlled quite successfully, still a large area of forest was affected by pests and disease every year. The young age structure and single tree species made it vulnerable to pests and disease. In addition, the number of forestry administrative punishment cases (it is a rough indicator to reflect the frequency of illegal logging) has significantly decreased after the 2005's forest tenure reform. Through redistributing the public benefit forest to individual farmer households, the reform created incentives for local farmers to manage and protect the public benefit forest. The farmers were encouraged to use the public benefit forest to develop agro-forestry and they became more active in protecting the public benefit forest. However, local economic development widened the

*Table 6.4. Public benefit forest changes in Liaoning Province (Liaoning Provincial Forestry Department, 2006b, 2009).*

| Year | Coverage | Annual growth in stock volume | Age structure[1] (%) |
|---|---|---|---|
| 2005 | 77% | 1.4% | 40:32:11:14:3 |
| 2008 | 79% | 5.4% | 44:48:4:3:1 |

[1] The age structure shows area proportion of stands at different ages including young, middle, pre-mature, mature, and over-mature.

Table 6.5. Main threats to public benefit forest in Liaoning Province (Liaoning Provincial Forestry Department, 2006b, 2009).

| Year | Forest fire | Forest pest and disease | | Illegal occupation | Law cases[1] |
|------|------------|------------------------|-----------|-----------|-----------|
| | affected area (ha) | affected area (ha) | treated rate | (ha) | |
| 2005 | 61 | 277,530 | 86% | 0 | 593 |
| 2008 | 81 | 255,351 | 95% | 13 | 197 |

[1] The law cases refer to forestry administrative punishment cases related to public benefit forests and most of them are illegal logging.

gap in revenue between public benefit forests and industrial development. It motivated illegal occupation on public benefit forests for mining and construction.

*Livelihood impact*

Since the forest tenure reform has been launched just after the installment of the payment scheme and the effects of the reform and the payment schemes intertwined with each other, it is difficult to quantitatively separate the impacts on the livelihood of farmer households of payment schemes and of the late reform. Therefore, this research assorted to the appraisal of local farmers for analyzing the impact of the payment schemes and used the data on income change to analyze the impacts of the tenure reform and its interaction with the payment schemes with regard to their implication on local livelihood.

Before the forest tenure reform, the public benefit forests in the sample villages were directly managed by the villages. The payment as a subsidy for public benefit forest management and protection only reached to villages, and villages selected forest rangers and their team leaders[36] with the approval of county forestry bureaus and paid for them using the subsidy. Although the forest rangers in principle should be village members, the amount of the payment was so little that the payment did not made a substantial impact on the overall livelihood of the villages (Table 6.6). Only those who have been employed as forest rangers have directly benefited from the payment schemes. This distribution method on the payment indeed strengthened the force of protecting the public benefit forest, but local farmers have been neglected as important stakeholders for effective forest management.

Moreover, the stringent implementation of public benefit forest protection imposed negative impact on the livelihood of local farmer households. According to the household survey, 54% of

[36] Public benefit management at township level is implemented by teams of forest rangers in Liaoning Province. In principle, one forest ranger takes charge of managing 200 hectares of forests and one team leader is in charge of monitoring and checking the work of ten forest rangers.

Table 6.6. Forest rangers and the payment of each village from 2001 to 2009.[1]

| Village | Population | Public benefit forest (ha) | Forest rangers | Overall payment (Yuan/year) |
|---|---|---|---|---|
| Yanghugou | 1,397 | 4,767 | 12 | 115,200 |
| Dayang | 1,075 | 1,200 | 6 | 57,600 |
| Beiwangqing | 1,380 | 2,309 | 4 | 38,400 |
| Luoquan | 621 | 1,133 | 3 | 28,800 |

[1] Source: interview with local village leaders.

the farmers thought that the payment schemes negatively impacted their income; 35% asserted the payment schemes had not affected their income; 7% believed their income had increased with the schemes[37]; and 4% had no idea on the impact on their income. When breaking down into each village, respondents indicating negative income effects by the schemes were not evenly distributed among the villages, with 79% in Luoquan, 62% in Yanghugou, 47% in Dayang and 25% in Beiwangqing. Luoquan is a typical village dependent on forestry production with the largest forestland area per capita (4.15 ha) and the smallest cropland area per capita (0.06 ha) among the four villages. In addition, its forest quality is highest with a stocking volume ranging from 105-120 cubic meters per hectare. Therefore, the farmers in Luoquan Village economically suffered most from the logging ban of the payment schemes. Likewise, Yanghugou Village has a similar natural resource structure for local livelihood, and 87% of its forestland has been included as a public benefit forest. As a result, the payment schemes had greater impacts on Luoquan and Yanghugou than on Dayang and Beiwangqing.

From 2005 to 2006, the collective forest tenure reform was carried out in the sample villages. The reform had an important impact on local economy and livelihood. The economic structure was increasingly reshaped, attributing to the redistribution of forest resources, the improvement on forestry investing environment and fast development on agro-forestry. According to the survey, the average household income increased by 48% from 11,144 Yuan in 2001 to 16,440 Yuan in 2009[38]. All sources of income have increased to a different degree after the payment schemes and the reform. When breaking down into each source of income, timber harvesting and agro-forestry showed a dramatic rise (Table 6.7). Before the tenure reform, both commercial and public benefit forests were managed by the villages and actually the power of management and decision making resided with the leaders of the villages. In principle, any decision on the management and transaction of collective forestland, and the use of the revenue generated from collective forests

[37] Some of them are the forest rangers and some are local farmers who succeeded in developing agro-forestry such as cultivating mushroom and Chinese herb medicine in forests.
[38] The monetary value of farmer household income in 2009 have been adjusted into real terms (taking 2001 as the base year), according to the consumption price index for rural residents provided by China Rural Statistics Yearbook 2010 (National Bureau of Statistics, 2010).

*Table 6.7. The change of average income and its structure from 2001 to 2009 (Yuan).[1]*

| Income | 2001 | 2009 | Increase rate | Share change |
|---|---|---|---|---|
| Total | 11,144 | 16,440 | 48% | – |
| Cropping | 2,887 | 3,131 | 8% | -7% |
| Animal husbandry | 506 | 756 | 50% | 0% |
| Off-farm work | 3,892 | 5,100 | 31% | -4% |
| Timber harvesting | 37 | 1960 | 51.97 | 12% |
| Agro-forestry | 1,006 | 2,113 | 110% | 4% |
| Small business and other | 2,817 | 3,382 | 20% | -5% |

[1] Source: the farmer household survey in Liaoning Province (n=54).

should be approved by village representative meetings or general members meetings according to the Organic Law of the Village Committee. Traditionally, in most of China's remote rural areas, political elites, including the director of village committees and the secretary of CPC at village level, usually ranks at the top of the power structure to control and manage village collective affairs. However, following the market-based economic reform and political reform of village elections and organization, economic elites (village entrepreneurs) start to challenge this power structure and obtain more advantage in the distribution of power (Chen, 2000). The shift in power structure also combines with the forest tenure reform. The emerging economic elites claim to limit or reduce discretionary power of village leaders on the management of collective forests. The redistribution of rural forest resources often faced opposition from local village leaders, who often had an interest in maintaining the status quo. In some villages in Xinbin County, the leaders stepped down under the pressure of the township government and village members because they opposed to redistribute profitable forestland to village members. The rapid jump in timber income was the result of redistributing commercial forests and the revenue flow now reaches the local farmers. However, timber harvesting is not the ultimate purpose of the reform. The provincial forestry officials argued that the collective forest tenure reform needs to lead to sustainable forest management. What makes it legitimate economically and environmentally is the subsequent investment and input in forestry after logging.

Besides the increase in timber income, the reform also triggered a substantial increase in agro-forestry – mainly cultivating Ginseng, wild vegetables, and Korean pine in public benefit forest. Different from the management paradigm on commercial forest, the reform adopted a new paradigm to manage public benefit forest, which aimed at easing conflicts between ecological conservation and local economic development. When the payment generated from outside (mainly from the central and provincial governments) was still quite low, this new channel – which takes the advantage of local natural environment improvements – offers additional incentive for protecting public benefit forest. In Liaoning Province, agro-forestry economy has evolved into an institutionalized practice to complement insufficient governmental payment under the extensive discourse of sustainable development.

Although average household income increased substantially, its structure changed in a different direction from before. On average, off-farm work still composes the main source of household income in 2009 (31%) with a decrease of 4%. The share of agricultural production in the average household income decreased by 7%. Compared to the decline in traditional agriculture and off-farm work, forestry related income experienced a fast rise and gained a large share in the household income (Table 6.7). The tenure reform and the ambitious plan for agro-forestry both contributed to the adjustment in local income structure. This change has a profound impact on the relationship between forest and local people. A new model of forest management is taking shape by restructuring rural economic activities with a built-in principle of ecological conservation. This process can be interpreted as ecological modernization in Liaoning's rural forested areas, where ecological rationality is increasingly changing forestry production.

Turning to the impact on daily life of local farmer households, we found that the forest use restrictions of the payment schemes also negatively impacted their daily fuel and other agricultural production, such as reduced supply of fuel and prohibition of livestock browsing on forestland. Over 57% of the respondents thought that the schemes reduced fuel wood consumption, 41% indicated that their family animal husbandry was affected, 2% mentioned other difficulties in their daily life due to the schemes and 26% of them asserted the payment schemes had no impact on their daily life (Figure 6.5).

## Participation

This section discusses how farmers were involved in the formulation and implementation of payment schemes and forest tenure reform. The analysis of farmers' participation not only focuses on policy formulation, demarcation, management and examination of the payment schemes, but also on the preparation and implementation of forest tenure reform.

a.  Payment policy formulation. Current payment policies have been developed at the national and provincial level. The process of policy formulation followed a traditional government-directed policy making paradigm. The payment schemes in Liaoning Province apparently

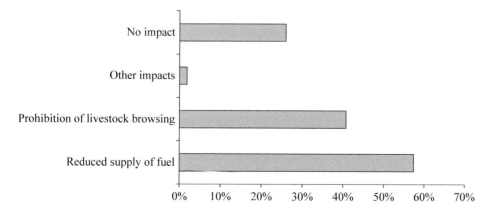

*Figure 6.5. Payment schemes' impact on daily life of local farmer household (n=54).*

have been shaped by two forces: the central government and local representatives of "the Two Meetings". The central government initiated the central payment scheme and provided both funding resources and an institutional base for the establishment of the provincial payment schemes. The local representatives also provided proposals during the "Two Meeting" of Liaoning Province to facilitate policy making on the payment schemes. For example, in 2008, 12 provincial People's Congress representatives promoted a proposal to the Meeting on establishing payment schemes for the water conservation area around the Dahuofang reservoir. This proposal was ranked first with the highest priority in the Meeting (in Chinese political culture, the first proposal means the highest priority for approval and implementation) and the provincial government responded to it seriously. In 2008, the provincial financial department promised to earmark 150 million Yuan over a number of years for paying ecological services in the eastern region – the major water conservation area, which also covers one of the sample counties – Xinbin. Hence, local representatives played an important role to render new payment schemes for ecological services. There is little evidence on direct involvement of local farmers at the stage of payment policy formulation. Some farmers still knew nothing about the payment policies of both central government (28%) and local governments (37%) after it had been implemented for more than 9 years (Figure 6.6), according to the survey. Lack of policy knowledge and lack of direct participatory channels for policy making caused a discrepancy between the current policy framework and the expectation of local farmers. Most apparently, the gap between the current payment standard and what farmers expected to accept is still large. The exclusion of villagers in policy processes that relate to their immediate interests such as stopping logging would promote an attitude of passivity and disengagement (Chia, 2009). Although the farmers accepted the environmental importance of forest ecosystem for local communities, they are reluctant to play a more active role in public benefit forest protection, such as tending forest to improve its quality and making a defense against illegal logging.

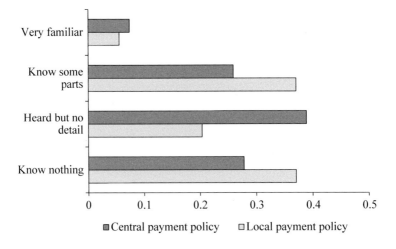

*Figure 6.6. Farmer's knowledge on payment policies (n=54).*

Referring to the question who has the responsibility for payments, 46% of respondents thought that the central government should solely take the responsibility; 9% of them believed that the central and local governments should be both responsible for the payment; 30% of them thought that the payment should be extracted from both the central government and the beneficiaries along the lower reaches of the watershed (residents and factories); 7% believed that three parties (including the central and local governments and the beneficiaries) should share the responsibility together. So in total, 92% of respondents thought that the central government should take part, although to a different degree, in the payment for public benefit forest. Actually, the local farmers understood the protection of public benefit forest, which mainly means stopping timber harvesting, as a duty of the central government, rather than of the local government or villages. At the same time, they believed that the central government should satisfy their basic needs for livelihood. If the central government fails in providing sufficient support for their living, it would be "legitimate" for them to start timber harvesting again. It should be noted that 37% of the respondents also related the paying responsibility to direct beneficiaries. However, this part is completely absent in current payment policies. Direct payment relationship between eco-service providers and beneficiaries is still missing.

b.  Demarcation of the public benefit forest. In 2001, the provincial forestry department sent out professionals to help county forest bureaus to develop the general layout of the public benefit forest according to the ecological importance of forestland. Updated satellite images and the data of forest resource inventories were employed to fix the locations and boundaries of public benefit forests at each forestland compartment. After demarcation, forestry bureaus informed the villages about forest land use change and made management contracts with village committees. The demarcation process was largely dominated by the provincial forestry department and county forestry bureaus and decided by scientific standards from the central and provincial forestry authorities. Although consultation of village members was carried out by village committees through all member meetings or representative meetings, local communities and farmers did not have any final decisive power on which plot of forestland can or should be included into public benefit forests. According to the interviews, local officials admitted that too much forestland had been demarcated into public benefit forests. The provincial forestry department standards are much more encompassing than the national standards, which usually have restrictions on unscrupulously including open forestland, shrub land, afforest land, and barren mountains. The controversy on the area of public benefit forest between the central government and the Liaoning provincial government continues to exist. In the 2009 Liaoning provincial report on public benefit forest management, the provincial forestry department still insisted that the central government should accept its demarcation result (22,666.67 $km^2$), but only 18,000 $km^2$ of its public benefit forest was confirmed by the central government and provided with payment (Liaoning Provincial Forestry Department, 2009). Although the contradiction on the standards reflects the self-interests of local governments in requesting for central funding, the central and provincial governments demonstrated high correspondence in their strong willingness to recover degrading forest ecosystems. Moreover, the demarcation process was relatively smooth in Liaoning Province, benefiting from collective tenure of forestland. Rather than negotiating with individual farmer households, the government only needed to reach agreement with village committees. According to the survey, 61% of the farmer households were consulted during the demarcation of public benefit forests. The forms of consultation included all-member meetings (26%[39]), representative meetings (37%), bulletin for public opinion (4%), and private negotiation (2%).

The farmers showed different degrees of acceptance of the demarcation result. 26% of the farmers welcomed the decision; 28% of them basically accepted it; and 46% of them opposed the change of forestland into public benefit forest. Among the opponents, the main reasons for disagreement related to the reduced supply of fuel and the failure in fulfilling contracts with villages (Figure 6.7). Before the payment schemes, villages had already contracted some plots of forest to local famers for timber production. These farmers were obliged to pay rent to the villages for the use of collective forestland and had to repay the loan if they already borrowed money from banks for forestry investment. Usually these farmers strongly opposed the change of the forestland into public benefit forest. Besides the individual investment, the requirement for fuel was a concern that resulted in opposition against the establishment of public benefit forest in rural area. Often daily energy is still dependent on local forest resources.

However, self-interest is not the only factor determining farmers' attitude to the payment schemes. When taking ecological importance of public benefit forest into consideration, more farmers (81%) showed support for the demarcation of public benefit forests. This also demonstrates that local farmers can separate environmental considerations from their private concern on income loss. In addition, most respondents (78%) thought that the process of demarcation was fair, in which each village and farmer household have been treated equally. The farmer household survey also showed that the farmers had different preferences for the process of demarcation: designation by governments (19%); designation by village committees

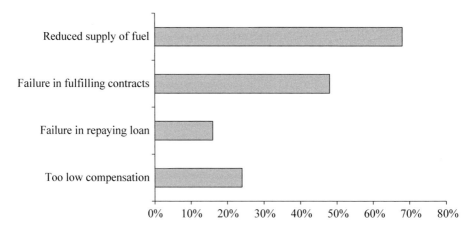

*Figure 6.7. Reasons for opposing the demarcation of public benefit forests (n=54).*

---

[39] It is the percentage of sample farmers informed and consulted by this form of consultation. And it is possible that farmers have been informed and consulted by multiple forms.

(28%); negotiation with individual households (5%)[40]; and designation by all-member village meeting (37%).

c.  Management of the public benefit forest. Management of public benefit forests in Benxi and Xinbin County is carried out by three administrative levels: the county, township and village level. The county public benefit forest management office takes charge of planning, coordinating, monitoring and examination on the management of public benefit forests, training of foresters, and the use, management and monitoring of the payment. The township stations for public benefit forest management are the main bodies for implementing the management and protection of public benefit forests. The village committees help to monitor public benefit forest management practices and provide county and township forestry authorities with the performance evaluation of forest rangers. After the county forestry bureau made an administrative responsibility contract with township governments, each village recommended local farmers for the position of forest rangers. Usually it is a competitive process since the position provides a stable job with relative high wage, compared to engaging in agricultural production. After the recommendation, the township stations for public benefit forest management examined the qualification of the candidates and some of them would be employed as forest rangers. The main task of forest rangers includes daily ranging in public benefit forests to prevent and control fire and pest; stopping illegal logging, occupying forestland, digging and burrowing, hunting, and clearing forest for cropping; reporting such illegal activities to local authorities; and other responsibilities mentioned in the forest management contracts. Under such administrative system, local farmers only participated in selection of forest rangers for public benefit forest. The survey showed that 69% of the farmers were consulted before deciding on employing forest rangers. 85% of farmers indicated that they completely (41%) or basically (44%) agreed on the choice of forest rangers by the village committee.

d.  Examination on the performance of the payment schemes. According to the Detailed Rules (Liaoning Provincial Forestry Department, 2006a), the examination on the performance of the payment schemes is carried out annually by municipal and county forestry bureaus together with financial bureaus. At first, county forestry bureaus examined the performance of each township and reported the result to municipal forestry bureaus. The municipal forestry bureaus verified the performance county by county and reported it to the provincial forestry department. At last the provincial forestry department organized a recheck on the performance of some selected counties. The content of examination includes the implementation of public benefit forest management (compliance with related regulations and laws, forest resources change, institution and organization building, information management, etc.), use of the payment, and the implementation of public benefit forest management contracts (ranging records, fire and pest control, preventing illegal logging and clearing forest, etc.). Although the Provincial Detailed Rules do not mention participation of local farmers in the examination, the counties developed more inclusive monitoring and management measures (*Benxi Manchu Autonomous County's Complementary Measures for Monitoring and Managing Public Benefit Forest Rangers*),

---

[40] It means that the village committees only negotiate with farmer households, which contracted forestland from villages and the land will be included as public benefit forest.

which to some degree included local farmers in the payment schemes. The county forestry bureaus set up fixed mailboxes in township forestry stations to collect public opinions on the performance of forest rangers in the management of public benefit forest. Moreover, the country public benefit forest Management Office selected 20 or 30 local farmers from each village (20 for relatively small villages and 30 for relatively large villages) for consultation about the performance of forest ranges. If the public opinion shows gross misconduct of one of the forest rangers or less than 60% of the local farmers selected are satisfied with his performance in public benefit forest management, he has to be dismissed from his post. According to the survey, 33% of the farmers had been consulted for opinions when forestry stations and high level governments conducted the (re)examination of public benefit forest management. In addition, 24% of the respondents have actively given opinions and revealed problems about the payment scheme implementation to the villages committees and the governments. Among them, 69% directed their input to township forestry stations, 8% to county forestry bureaus and 54% to village committees[41]. Township forestry stations and village committees are the main channels to absorb local farmers' opinion on public benefit forest management. The survey showed that efficiency of performance examination was considered highest with forestry stations, and less with monitoring of local farmers, with county forestry bureaus, village committees or provincial forestry departments (Figure 6.8).

The results showed that existing mechanisms of monitoring and examination are basically corresponding with the expectation of farmers. Township forestry stations and county forestry bureaus play a pivot role in examination with a substantial involvement of local farmers for monitoring. Formal and informal rules for local farmer's participation in the examination, especially for monitoring the performance of forest rangers, resulted in a frequent interaction

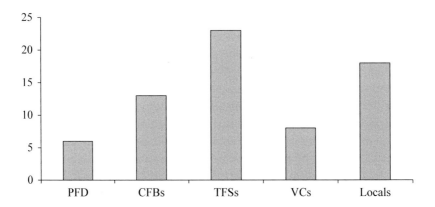

*Figure 6.8. Frequency of preference to public benefit forest examination body (n=54).*
*PFD: provincial forestry department; CFBs: county forestry bureaus; TFSs: township forestry stations;*
*VCs: village committees; Locals: local farmers.*

---

[41] Some farmers directed their opinions to village committees, forestry station, and bureaus together.

between local farmers and township forestry stations and county forestry bureaus. This interaction helped farmers to understand the responsibility of forestry authorities at different levels and improved efficiency of the public benefit forest management. It also increases legitimacy of the role of forest authorities.

e. Participation in collective tenure reform on public benefit forests. The collective forest tenure reform as a comprehensive institutional change, includes different components and implementing stages: redistributing forest plots to farmers (clarifying boundaries of forest plots, contracting out plots and issuing certificates of forest tenure), establishing supporting policies (forest tenure exchange market, forest tenure for loan policy, forest insurance policy, disclosure policy on forest logging information, forestry association or cooperation policy, etc.), and a reform of public benefit forest management and protection mechanism reform. To be specific, we focus more on the distribution of forest plots to farmers, which is the principal part of the reform.

The central and local governments have emphasized participation of local farmers during the reform (Liaoning Provincial Government, 2005a). At first, each village established a working team for the reform, which thoroughly investigated forest resources of the village and the preference of villagers on the apportionment of collective forest among them. The core part of the reform is making an implementation arrangement on how to distribute collective forestland and divide corresponding rights and obligations. The draft implementation arrangement should be agreed upon by at least two third of the villagers or village representatives. After the villagers accept the arrangement, it is sent to the township government for approval. Then the village working team confirms and registers tenure, boundaries and areas of each parcel of forestland and makes a contract with farmer households according to the arrangement. The inventory and contracts are sent to township governments for checking. At last, the municipal and county forestry bureaus examine the performance of the reform. According to the survey, 94% of the respondents knew quite well or basically the forest tenure reform arrangement of the villages. There is also a high agreement rate in the villages, with 94% of them being satisfied. After the reform, 92% of the respondents received formal forest tenure certificates issued by county forestry bureaus. Furthermore, 86% of them thought the reform significantly or basically clarified and legalized the forest tenure of the villages. In addition, 98% of respondents supported the redistribution of tenure of the public benefit forest from villages to farmer households. The main reasons for their support included facilitation of the payment distribution[42], the improvement of public benefit forest protection, and the development of agro-forestry (Figure 6.9).

f. Participation in ongoing management reform. Redistributing forest to farmers caused a problem on the legitimacy of the payment schemes since the public benefit forest contracts were sign only between the country forestry bureaus and the village committees. After the tenure reform, the de facto owners of the public benefit forest are local farmers. Therefore, a new contract had to be introduced between these new owners and the forestry authorities. Furthermore, the increasing revenue gap between commercial and public benefit forests, which

---

[42] The tenure reform clarified and substantiated the ownership of public benefit forest. It is expected to guarantee that individual farmer households as public benefit forest owners can obtain the payment.

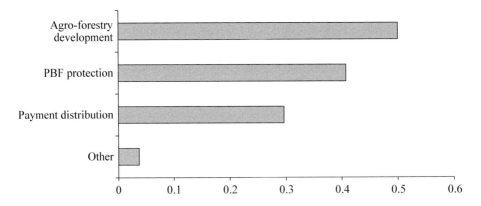

*Figure 6.9. Reasons for supporting public benefit forest (PBF) tenure redistribution (n=54).*

is induced by the forest tenure reform, puts the management and protection of public benefit forests at risk. It became imperative to clarify the obligations and rights for public benefit forest owners. In 2009, the Liaoning provincial forestry department issued *the Opinion on Facilitating Reform on the Management and Protection Mechanism on Public Benefit Forest* and formally started a provincial wide reform on the public benefit forest management. The reform had a two year plan to settle down the responsibility of the public benefit forest by various contracts with local forest owners including professional management (collective public benefit forests not contracted out), entrusted management (less forest-dependent regions), cooperative household management (heavily forest-dependent regions), and individual household management (large area forest owners). The reform also aimed at channeling part of the payment to individual farmer households directly. This part of the payment includes replanting and tendering subsidy from the original payment (22.5 Yuan per hectare) and the additional part of the new payment policy (75 Yuan per hectare) since 2009. Since the reform is still going on, only one-third of farmer households have made management contracts with the county forestry bureaus, according to the survey.

However, as mentioned above, most farmers were dissatisfied with current payment standards and increasing the payment is imperative to ensure the success of management reform of public benefit forests. According to the survey, the payment standard, even the new one for 2010 (150 Yuan per ha), is still significantly lower than the expectations of local farmers (Figure 6.10). The survey also included questions on choice for alternative policies for public benefit forest management. The first alternative is to establish a biding scheme to decide which plot of forest is included into public benefit forests with how much payment on a voluntary basis. In this scheme, villages and farmer households initiate payments for plots of forest. The government chooses forest plots as public benefit forest based on the payment required, ecological importance, and forest resources quality. After explanation of the mechanism of this bidding scheme, 63% of the respondents thought that this would be a better way of public benefit forest management than the current way; 9% of the respondents opposed it; 28% of them did not know. The second alternative is to adjust some parts of a public benefit forest into a commercial forest or the opposite. No one was interested to turn their commercial forest into

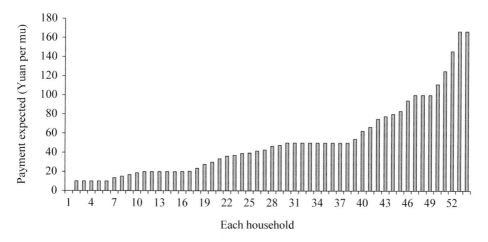

*Figure 6.10. Farmers' expectations to payment (n=54).*

a public benefit forest; 46% of the respondents wanted to change some of their public benefit forest into a commercial forest; 54% of them preferred to make no adjustment to their public benefit forest.

## 6.5 Conclusion

Through the implementation of payment schemes, the idea of PES has been built in local sustainable forest management in China. However, it is different from the PES in other developing countries such as Costa Rica, in terms of discourse, actors involved, rules and power relations. Unlike a neoliberal paradigm relying on the power of market, China's payment schemes for forest ecological services mainly rest on scientific planning, national subsidy, and political will. Farmers regard PES as a direct exchange with the state, in which they simply stop timber harvesting and the state guarantees their basic livelihood requirements. This consensus is quite different from the market-based PES in which a voluntary exchange is based on ecological value generated from the forest management. Likewise, direct beneficiaries, such as factories and residents along the lower reaches of watersheds, often have not been included into the payment schemes. Although several governmental documents mentioned it is necessary to include these important stakeholders into PES schemes (Liaoning Provincial CCP Committee and Liaoning Provincial Government, 2009), the effort to do so often fails as it faces strong opposition from interest groups of these sectors. As a result, the payment schemes have been designed as single-sector narrow policies without integration with other sector policies. It is set up centrally and implemented top-down. The central government is the main coordinating, funding and implementing agency of forest payment schemes. However, following the institutionalization of the central payment scheme, provincial and local forest payment schemes have emerged more recently. This was also facilitated by the forest tenure reform which shifts the tenure from villages to farmer households. The ownership change facilitates the legitimacy of farmers' request on compensation and causes the governments to create rules and funding resources to promote agro-forestry in public benefit forests in order to ease the conflicts between nature conservation and livelihood impact.

The institutional setting in Liaoning Province provides important support for implementation of payment schemes and guarantees their performance. The setting consists of 4 pillars: classification-based forest management, target responsibility system, forest tenure reform, and forestry industrial policy. Classification-based forest management is written into the Forest Law as a legal basis for delimiting and managing public benefit forest. Target responsibility system is an institutional support for monitoring and examining the implementation of payment schemes in current administration system. The forest tenure reform in Liaoning Province is establishing a stable tenure structure for public benefit forests, which is a precondition for well-functioning payment schemes and opens an opportunity for involvement of market actors in the future. Forestry industrial policy as the last pillar coordinates ecological conservation and rural development in forested areas. It provides both rules and funding for developing agro-forestry in public benefit forests.

On the other hand payment schemes also have influenced the institutional setting in Liaoning province. The practice of payment schemes pushed for adjusting classification-based forest management in Liaoning Province, which changed from relying on "command and control" measures to emphasizing economic incentives for forest protection. The implementation of payment schemes induced a policy debate on whether public benefit forests should be involved into forest tenure reform. In the end, the forest tenure reform included public benefit forests and largely reconstructed the tenure structure of the public benefit forest in Liaoning Province. The inclusion of public benefit forest protection and management into target responsibility system can be regarded as a centripetal move of ecological rationality into the current administrative system. In addition, the payment schemes induced the transformation of forestry industrial policy from focusing on timber production to coupling forestry development with ecological conservation.

While payment schemes interact with the institutional setting at different levels, they are influencing local forest use practices in a different way and to different degrees in distinct communities. In general, more ecological improvements have been achieved in the regions which are more fragile to soil erosion. More efforts in forest protection are put in these fragile regions. The more the traditional economic structure of the community depends on extracting forest resources, the more the livelihood of the communities suffers from the payment schemes. Therefore, the risk to revert to previous forest use practices is high in these communities and also the cost of monitoring and coordination is relative high compared to less forest dependent communities. In addition, if a community can utilize its forest resources to develop agro-forestry and thus change its economic structure, the payment schemes will have less negative effects on local livelihoods and the forest use practices will more smoothly and rapidly change from timber extraction to coupling agro-forestry with ecological conservation.

In understanding PES in Liaoning Province, the assessment showed that the effects of the payment schemes on environmental services are most likely positive. After the forest tenure reform settles down the forest ownership, forest owners even become more active in participation in forest protection and restoration. However, the age structure and tree species composition is still in a poor condition and current payment schemes did not provide an efficient mechanism to deal with it.

Payment schemes had a negative impact on the income of local farmer households, especially in regions where the public benefit forest had been used previously for timber production. The negative impact on local livelihood truly has effects on the attitude of local farmers towards the payment schemes and it could lower the effectiveness of the schemes if the income loss of the farmers cannot be fully compensated by the payment or other channels. After the forest tenure reform, agro-forestry generated additional income to ameliorate the livelihood of local farmers. This positive effect also guarantees maintaining the effectiveness of the payment schemes.

The central and provincial governments still play a key role in initiating, designing, implementing, managing and examining payment schemes. Local forest owners/users have participated in the implementation of the schemes, including their demarcation, management and examination. During the forest tenure reform, local farmers gained more opportunities for direct participation. The reform also created a legitimate position for farmers to participate in public benefit forest management. This positive change in participation contributed to further improvement of the livelihood of local farmers through both the forest tenure reform and the payment schemes. In addition, the investigation showed that even with similar negative economic impacts on local income, broad and sufficient consultation with local farmers increased their support for the payment schemes. The support can guarantee to a large extent compliance to the restrictions imposed by the schemes. However, channels for participation of local farmers are still needed to extend to effectively communicate with them, especially in the demarcation of the public benefit forest.

Overall, the implementation of payment schemes for forest ecological services in Liaoning Province shows improvements in forest management and protection, as it not only breaks away from previous single minded timber extraction but also increasingly employs an incentive-based governance method instead of "command and control". The payment schemes created new forest use practices which couple local livelihood requirements with ecological conservation. Furthermore, local farmers are more and more involved in the implementation of payment schemes, rather than marginalized consumers of decision regarding their traditional sustenance sources.

# Chapter 7.
# Conclusion

## 7.1 Introduction

During the last quarter of the twentieth century China exploited its forests in fueling its economic boom. Deforestation for wood harvesting, agricultural development and energy development came together with ineffective forest protection policies. Not only did the forest area decrease, also the ecological quality of the forest deteriorated. But China's forest policies have changed tremendously since 1998. Most importantly, the devastating flood along the Yangtze River sounded an alarm for the enduring environmental degradation and depletion of forests in the country. The flooding proved that the regulation on forest management in the past decades had failed to meet its objective of providing a sustainable forest ecosystem. In addition, the increasing fiscal revenue following the 1994 Fiscal Reform strengthened the central government's financial ability to provide pubic goods, including ecological restoration and environmental protection. As such, over the past decade forest ecosystem protection has changed from mainly regulating and controlling forest logging to employing a variety of new financial instruments to motivate a variety of actors into forest protection and restoration. This can be interpreted as an example of political modernization of forest policies.

Since 1998, China has launched six national forest ecological conservation and restoration programs to combat forest destruction and to reduce negative impact on forest ecosystems in China. Funding from the central government has focused on three programs[43]: the Natural Forest Protection Program (NFPP), the Conversion of Cropland into Forest and Grassland Program (CCFGP), and the Forest Ecological Benefit Compensation Fund Program (FEBCFP). Planned investment for NFPP and CCFGP has exceeded 316.2 billion Yuan (US$ 45.2 billion) in total by 2010, and FECFP has an annual budget up to 3 billion Yuan (US$ 428.6 million) (Zhou, 2001). With such a huge capital investment in these programs, questions have arisen about the environmental effectiveness, efficiency and socio-economic consequences of the implementation of these programs, especially compared with conventional "command and control" policies. The academic community and governments have carried out a flurry of evaluations of especially NFPP and CCFGP, with little attention to the FEBCFP. Part of the reason for this imbalance in forest policy evaluation is that in the short term, NFPP and CCFGP provided a larger capital flow than FEBCFP and therefore plausibly had more significant environmental and socio-economic consequences.

Results from various policy evaluations have shown that NFPP and CCFGP initially gave a strong punch to local forest industry and agricultural economy and produced positive environmental consequences (State Forestry Administration, 2006b; Xu *et al.*, 2004, 2006a). However, these evaluations also criticized the unsustainability of the environmental achievements, since the

---

[43] The three other programs were Sandification Control Program in the Vicinity of Beijing and Tianjin (SCPVBT), Shelterbelt Program in Three North area and along Yangtze River (SBP), Wildlife Conservation and Nature Reserve Development Program (WCNRDP).

programs solely depended on short term national investment[44] without any formal payment institution for the period of post-NFPP or CCFGP (Uchida *et al.*, 2005; Xu *et al.*, 2006a; Zhang *et al.*, 2008). Compared to the NFPP and the CCFGP, the FEBCFP has a more stable financial source. It is supported by regular budgetary revenues, covering almost the whole country. And, in contrast to the national revenues of the other two, the FEBCFP succeeded in involving provincial and local governments to develop their own payment schemes for forest ecosystem services. In 2009, 25 provinces had established their own provincial payment schemes for forest ecological services (State Forestry Administration, 2010a). Although this looks promising, few empirical studies have evaluated the implementation and performance of these more regional and local payment schemes for forest services. This study has set out to provide more insight in the functioning of these payment schemes in the Chinese forest management sector.

In order to assess the performance of payments schemes for forest ecosystem services, an evaluative framework was developed, including three aspects: environmental effectiveness, economic impacts and participation. This analytic and evaluative framework aimed to answer the following research questions:

1. What have been the ecological and socio-economic effects of forest PES schemes in China?
2. To what extent and how have state and non-state actors (including farmers) participated in the design, implementation and evaluation of forest PES schemes in China?
3. How has forest tenure reform influenced the functioning and outcome of forest PES schemes in China?

This study used a case study approach in three provinces to assess the functioning of payment for forest ecosystem services in China. This chapter will draw the conclusions of that research. In the next section I will start with comparing the case studies from Liaoning, Guangxi and Fujian provinces in order to formulate answers to the research questions. Section 7.3 gives reflections on the research, especially on the evaluative framework, research methods and theory. Finally, Sections 7.4 and 7.5 provide policy recommendations and suggestions for future research.

## 7.2 Comparison of the cases

This section will explore what the three cases have in common, in what ways they differ in terms of the research questions, and what general conclusions can be drawn out of these cases.

### 7.2.1 Various payment schemes

In each province or autonomous region the research has investigated two or three payment schemes for forest ecological services (see Table 7.1). The central payment scheme has been implemented in all three provinces/autonomous regions and funded by the central government. It mainly covers national public benefit forest, but it also supports regional public benefit forest (PBF) for poor regions such as Guangxi Autonomous Region. The implementation of the central

---

[44] The duration of the NFPP is 10 years and that of the CCFGP was originally 8 years for forest regeneration but has been extended recently to 16 years.

Table 7.1. Comparison of payment schemes in the three cases.

| Cases | Current schemes | Implementing organization | Funding | Coverage[1] | Main usage |
|-------|-----------------|---------------------------|---------|-------------|------------|
| Fujian | central forest ecological benefit compensation fund | provincial forestry departments or regional forestry bureau | central government | national PBF | compensation for forest owners, |
| | Fujian provincial forest ecological benefit compensation fund | | provincial government | provincial PBF | hiring |
| | Min River watershed forest ecological compensation fund from downstream to upstream regions | | provincial and municipal governments | all PBF | foresters, pest and fire control, capacity |
| Guangxi | central forest ecological benefit compensation fund | | central government | national and regional PBF | building for PBF management |
| | Guangxi Regional forest ecological benefit compensation fund | | regional government | regional PBF | |
| Liaoning | central forest ecological benefit compensation fund | | central government | national PBF | |
| | Liaoning provincial forest ecological benefit compensation fund | | provincial government | provincial PBF | |
| | Liaoning provincial natural forest protection compensation program | | provincial government | natural forest | |

[1] PBS: public benefit forest.

payment scheme involves the State Forestry Administration, the Ministry of Finance, provincial forestry departments and financial departments, county forestry bureaus and financial bureaus, and township forestry stations. But provincial forestry departments play a central role in managing the payment schemes, including distributing funds, demarcating PBF, and the monitoring and examination of the performance of PBF management and protection. The payment from the central scheme is generally used for PBF management and protection (including hiring, training, and organizing local foresters, and implementing fire and pest control) and compensation for forest owners.

The payment schemes at the provincial or regional level also have been developed in each province or autonomous region, according to the requirement of the central government. These provincial payment schemes are completely funded by the provincial or regional governments and they only cover PBFs at the provincial level. Like the central payment scheme, the provincial schemes are implemented by the provincial forestry departments or the regional forestry bureau, which also follow the same rules and procedures as the central schemes.

Among the three cases, Fujian and Liaoning have their unique payment schemes, which, in several ways, differ from those provincial payment schemes mentioned above. These payment schemes have been established by the provincial governments voluntarily and functioned as complements to the national and provincial payment schemes. Furthermore, the watershed compensation scheme in Fujian has been designed as a compensation for forest owners rather than just a subsidy to forest management and protection. In addition, these payment schemes developed a more flexible funding mechanism, which pooled money from government budgets at different levels. This mechanism de facto puts more funding responsibility on regions, which benefit more from forest ecosystem services, and delivers more financial resources to regions, where more forest resources have been taken away for ecological purpose.

### 7.2.2 Environmental effectiveness

The three cases show different trends in managing forest resources, improving forest ecosystem services and reducing soil erosion. Through both incentive mechanisms and regulatory methods, the payment schemes changed forest use practices from timber production to ecological conservation. Among the three cases, Liaoning's payment schemes covered the largest areas of its forest land (Table 7.2). Fujian just reached the national requirement for public benefit forests as stipulated in the Forest Law (30% of forestland should be designated as public benefit forest). Due to limitations on forest quality data, it is hard to directly compare the improvements among three cases by a set of uniform standards. But forest inventory data on sampled forest plots and provincial public benefit forest resources can give a general trend of forest quality. According to these results, Fujian achieved most in improving forest quality for ecological services, with increases in all quality indicators. Comparatively, the public benefit forest in Guangxi did not show significant change in its quality. Although Liaoning slightly improved its public benefit forest in terms of stocking volume and coverage, the forest structure is still relatively poor with a low proportion of mature forests, which implies low capacity of ecological services. The assessment from the farmers showed consistent observation to the forest investigation data. In Fujian and Guangxi, 60-70% of the respondents confirmed that the implementation of payment schemes already to some degree reduced local soil erosion. However, only 40% of respondents in Guangxi observed slight reduction of soil erosion. In addition, a relatively high proportion of Fujian's respondents asserted a significant reduction on soil erosion. As a result, we can rank the environmental effectiveness of payment schemes in the three cases as follows: Fujian, Liaoning, and Guangxi (see Table 7.2).

There are several factors that cause this difference in environmental performance of the payment schemes in different cases. First, institutional factors determined how the payment schemes had been formulated and implemented and further had an impact on the environmental improvements. During the investigation, Fujian and Liaoning have completed their first stage of collective forest tenure reform – clarifying boundaries and issuing forest tenure certificates. The tenure reform clarified the rights and responsibilities of local farmers in public benefit forest management, and increased their involvement in the management of forest resources, which were regarded as common pool resources without effective management and protection in the past. In

*Table 7.2. Performance of different cases on environmental effectiveness.*

| Cases | Forest investigation data | | Farmers' assessment on soil erosion[3] | | |
| --- | --- | --- | --- | --- | --- |
| | protected forest land[1] | forest quality[2] | significantly reduced | slightly reduced | in total |
| Fujian | 30.7% | significantly improved | 35% | 29% | 64% |
| Guangxi | 34.8% | no significant change | 0% | 40% | 40% |
| Liaoning | 41.8% | slightly improved | 24% | 46% | 70% |

[1] Percentage of public benefit forest in all forestland.

[2] The assessment on forest quality based on forest inventory data provided by local forestry bureaus.

[3] Farmer's assessment on soil erosion prevention is based on farmer household survey carried out in the three cases.

contrast, Guangxi's tenure reform was still in preparation and had not provided a motivation for improving public benefit forest protection (see Table 7.3).

Furthermore, political willingness of governments for initiating and implementing payment schemes is also an important factor. It determines how many resources (funding and staff) the provincial governments can contribute to the implementation of the payment schemes (including fire and pest control, frequency of examination, forest data management and training for foresters). Political willingness to develop payment schemes also pushes the formulation of local rules and procedures for public benefit forest management. Both Liaoning and Fujian provincial

*Table 7.3. Factors for the difference in environmental effectiveness.*

| Category | Factors | Cases | Impacts |
| --- | --- | --- | --- |
| Institutional | collective forest tenure reform | Fujian, Liaoning | positive |
| | political willingness | Fujian, Liaoning | positive |
| | administrative status | Fujian, Liaoning | positive |
| | forestry industry policy | Liaoning, Guangxi | positive for Liaoning, negative for Guangxi |
| Payment schemes | high payment standards | Fujian | positive |
| | diversity of payment schemes | Fujian, Liaoning | positive |
| | sufficient consultation | Fujian, Liaoning | positive |
| | local implementation measures | Fujian, Liaoning | positive |
| Contextual | dependence on forest resources | Guangxi, Fujian | negative |
| | alternative livelihood opportunity | Fujian, Guangxi | positive |
| | natural disaster | Guangxi | negative |

governments mobilized funds for developing provincial payment schemes and other local schemes. Fujian provincial government already set up a specialized office for coordinating payment schemes. Likewise, Liaoning province established training programs for foresters to improve public benefit forest management and developed a set of strict monitoring and examination measures. Political willingness not only reflects the support from provincial governments, but also relates to the administrative status of forestry departments and their capacity to motivate their provincial governments. In general, a higher status of forestry departments in the administrative system means having a relatively strong influence of that department on the political agenda of the province and a high capacity to mobilize resources for forestry from provincial governments. This factor usually combines with political willingness to influence the implementation of payment schemes. Different from other institutional factors, forest industry policy has profound impacts on the environmental performance of payment schemes through providing positive or negative incentives for public benefit forest protection. Positive forest industry policies, like those in Liaoning, take both forest industry development and public benefit forest into account, rather than only emphasizing commercial forest production like Guangxi's forest industry policies. This implies the combination of ecological rationality with economic rationality in forestry production and forest protection. Liaoning is an excellent example for this process of ecological modernization in its forest domain – ecologically improving its forest industry and economically managing its forest ecosystem.

Second, the characteristics of payment schemes have a direct influence on their environmental performance. The foremost factor is the compensation standard on public benefit forests. Among the three cases Fujian has the highest standard for paying local farmers. Although it was still lower than the expectation of local farmers, the schemes to some degree generated incentives for farmers to change their forest use from timber production to ecological conservation. In contrast, the degree of compliance of local farmers in Guangxi was lower than those in Fujian and the conflicts between local livelihood and public benefit forest protection was more apparent. Furthermore, besides the central payment schemes, each province or region has developed its provincial payment schemes. Fujian and Liaoning initiated new payment schemes through different funding and these payment schemes complimented the central payment schemes with additional payment, paying channels and extended forest areas. Although the central government stipulated the principle of voluntariness in the demarcation of public benefit forests, local governments had significant room to use their power to "force" farmers to collaborate in practice. As a result, the degree of participation of local farmers in the payment schemes differed from region to region. The survey results show that the involvement of local people can increase their acceptance of the payment schemes, even with current low payments. Especially during the stage of demarcation of public benefit forests, consultation with local farmers can avoid conflicts between forestry authorities and local people, increase legitimacy of payment schemes and improve their environmental effectiveness in the end. In addition, although the central government developed general implementation measures for payment schemes, some provincial and county governments had tailored the measures to meet local conditions (forest resources, local livelihood, and forest use practices). In Fujian and Liaoning, every county has developed such implementation measures and dynamically improved those measures according to contextual and institutional changes (such as forest tenure reform).

Third, contextual factors also played an important role in achieving environmental effectiveness. Natural conditions, characteristics of the community and the local economic and social situation form a context for the functioning of payment schemes. Among these contextual factors, dependence on forest resources for living is the most important factor that hinders the acceptance by local farmers of payment schemes and thereby negatively impacts on environmental performance. It is apparent in the Guangxi's case that traditional dependence on forestry production for living (timber harvesting and bamboo plantation) caused a relatively strong opposition from local farmers to the payment schemes. Even in the same county, the dependence on forestry production varies from village to village. Under the current design of a uniform payment standard, this variety makes it difficult for governments to optimize payments to local farmers. However, the dependence on forest production is not the only decisive factor that influences local farmers' behavior in forest use. Whether an alternative livelihood is available can also determine the ease of changing current forest use practices into more sustainable ones. This opportunity for alternative livelihood is related to local natural landscape, forest resources, market access and information, and the potential for developing tourism, agro-forestry and other non-timber forestry production. Of course, governments can provide important support for creating new opportunities and boosting ecotourism and agro-forestry in a sustainable way. For example, Liaoning forestry departments employed forest tenure reform and forest industry policy to facilitate the transformation in forest use practices based on local natural and forest resources conditions. Profitable models for developing agro-forestry have been introduced to village communities and at the same time, governmental support has been provided for local farmers. In addition, natural disasters, such as a severe snow storm, are uncontrollable factors determining the environmental performance of payment schemes. The 2008 heavy snow storm destroyed the public benefit forest located in fragile regions of Guangxi and reduced its capacity for ecological services significantly. This is also an important reason for the low environmental performance of payment schemes in Guangxi.

### 7.2.3 Economic impacts

The case studies showed that payment schemes achieved different environmental effectiveness in all three cases. The variation in environmental performance closely relates to the impacts on local livelihood and the distribution of cost and benefits among local farmers. Results from all three cases showed that payment schemes impacted the income of local farmers (Table 7.4). Farmers' livelihood in Guangxi has been the most heavily stricken by the payment schemes among the three cases. 89% of the respondents in Guangxi (the highest proportion in all cases) asserted negative impacts of the schemes on their income. In contrast, most of the farmers in the Fujian case thought that the payment had no apparent negative impacts on their income. The response of farmers in Liaoning was between those of Fujian and Guangxi, where about half of the respondents indicated negative income impacts of the payment schemes.

Forest resources not only generate income from timber harvesting for rural communities but also support other types of non-forestry production. Apart from the impact on income, the implementation of payment schemes also had negative impacts on raising local livestock and fuel consumption (Table 7.5). Local farmers (especially in Guangxi, where daily energy consumption is largely dependent on firewood from forests around the communities) complained that the

*Table 7.4. Impacts of payment schemes on household income.*

| Cases | Farmers' assessment on income impact | | | |
|---|---|---|---|---|
| | negative | positive | no impact | don't know |
| Fujian | 14% | 10% | 71% | 5% |
| Guangxi | 89% | 11% | 0% | 0% |
| Liaoning | 54% | 7% | 35% | 4% |

*Table 7.5. Percentage of farmers that experienced impacts of payment schemes on livestock raising and fuel wood consumption.*

| Cases | Reduction of fuel wood | Prohibition of livestock browsing |
|---|---|---|
| Fujian | 33% | 14% |
| Guangxi | 68% | 21% |
| Liaoning | 57% | 41% |

establishment of payment schemes reduced their access to local forest resources and reduced their traditional fuel supply. In Fujian, the negative impacts were smaller than those of the other two cases. This was thanks to a recent transformation of rural energy consumption from firewood to coal and a fast growing diffusion of household-based digesters in rural areas. Furthermore, prohibition of livestock browsing in public benefit forests influenced and terminated traditional animal husbandry practices. This negative impact was more apparent in northern regions such as Liaoning, where raising livestock relies more on browsing in the natural environment, including forests.

Since the start of the payment schemes the income of local farmer households has changed along totally different trajectories and its source structure has shifted into different development models (Table 7.6). In Guangxi, the income of local farmer households experienced a large decline (going down by 36.6%) since the payment schemes began. During the same period, the income of Fujian's local farmer households showed a minor fall (by 7.0%). However, the income of Liaoning's farmer households surprisingly increased by 47.5% from 2001 to 2009. The three cases demonstrated quite different income changes during almost the same period. The payment schemes are pivotal factors to induce such changes and apparently they played different roles in the three cases. But other factors, such as the institutional setting and the local context, also to some degree differentiated the impacts of the payment schemes.

Together with the income change, the shares of various income sources shifted in different directions. In Guangxi, timber production has been heavily inflicted and its leading position in

*Table 7.6. Local farmer household income change during the payment schemes.*

| Cases | Period | Income increase (%) | Income structure[1] | | | | | |
| --- | --- | --- | --- | --- | --- | --- | --- | --- |
| | | | Timber harvesting | | Off-farm work | | Agro-forestry | |
| | | | before[2] | after[3] | before | after | before | after |
| Fujian | 2001-2007 | -7.0% | 1.2% | 0.0% | 52.0% | 51.0% | 0.0% | 0.0% |
| Guangxi | 2001-2007 | -36.6% | 60.0% | 17.5% | 13.3% | 41.1% | 0.0% | 0.0% |
| Liaoning | 2001-2009 | 47.5% | 0.3% | 11.9% | 34.9% | 31.0% | 9.0% | 12.8% |

[1] Only forestry related income sources and off-farm work are presented.
[2] In the year 2001 for all cases.
[3] For Fujian and Guangxi, 2007 income data were recorded; for Liaoning, 2009 data were collected in order to analyze the impacts of collective forest tenure reform.

income contribution has been replaced by off-farm work. It implies that in the implementation of payment schemes in Guangxi, local farmers have lost most of their access to forest resources around their communities, which fundamentally changed their living practices. However, the Fujian case showed that income structure remained the same. Although timber harvesting was prohibited, it did not change the income source structure of the farmer households dramatically, since timber harvesting was not a major source for their income before the payment schemes. For Liaoning, the story is quite different. There the contribution of timber harvesting and agro-forestry to household income has increased significantly, though off-farm work has kept its leading position as an income source, be it with a relatively slight decline.

It is certain that the implementation of payment schemes has imposed direct impacts on the income of local farmer households. However, interestingly each case showed a different type of change in the income of farmer households. It is justifiable to look for other factors from the institutional setting and local context which impact on household income and livelihood (Table 7.7). Each of these factors certainly does not function alone, but rather the different factors interact with each other and together influence the local rural economic activities and forest use practices.

First, payment schemes brought various forest management measures into public benefit forests. The main factors influencing local livelihood include logging restrictions, payments, demarcation of public benefit forests, and alternative forest use. Payment schemes imposed logging limitations on public benefit forests, which universally changed local forest use practices. In all three cases, the restriction on timber harvesting had negative impacts on the income of local farmer households. Especially in Guangxi, where timber harvesting had been the major income source, local farmer households suffered severe income decline from logging restrictions on the public benefit forest. Furthermore, payment schemes have been designed to reduce the negative impact of logging restrictions and to facilitate the implementation of public benefit forests.

*Table 7.7. Factors impacting local household income and livelihood.*

| Category | Factors | Cases | Impacts |
|---|---|---|---|
| Institutional | collective forest tenure reform | Liaoning, Fujian | positive |
| | forestry industry policy | Liaoning | positive |
| Payment schemes | logging limitation | Fujian, Guangxi, Liaoning | negative |
| | payments | Fujian, Guangxi, Liaoning | positive |
| | arbitrary demarcation of pbf | Guangxi | negative |
| | sustainable forest management | Fujian, Liaoning | positive |
| Contextual | dependence on forest resource | Guangxi, Liaoning | negative |
| | alternative livelihood opportunity | Guangxi, Liaoning | positive |

Although payments distributed by the schemes have to some degree ameliorated income loss of local farmer households, most of the respondents thought it was too low to fully compensate their loss from the schemes. Therefore, the payments functioned as a subsidy for strengthening public benefit forest management rather than a compensation for income loss. However, the nature of payments, whether it is a subsidy for forest management or a compensation for income loss, also varies from case to case, depending on how the payments have been distributed among government, villages and local farmers. In Liaoning, all the payments have been used to organize and hire foresters to protect public benefit forests, who were also selected from local farmers. Under such circumstances, the payment is used more like a subsidy with regulatory methods on forest protection. In contrast, the payments of the compensation scheme from downstream to upstream regions in Fujian have been directly channeled to local farmers to encourage them to change forest use practices and reduce timber harvesting. In this situation, this kind of payment works like a compensation to reduce conflicts and create incentives for the transformation of forest utilization. Other factors in the implementation process of payment schemes also influence the local livelihood. In principle, the demarcation of public benefit forests should take place in consultation with local communities and farmers. However, in practice, the demarcation has been carried out by forestry authorities unilaterally and the villages and local farmers have been only informed of the result after all demarcation was done. Low levels of participation of local communities in the process of demarcation caused that concerns of local people on local forest-based livelihoods have been neglected in the name of uniform planning. As a result, in some villages (typical in Guangxi) all forests – both natural and plantation – were designated as public benefit forests. This brought a radical change to local livelihood, especially in the regions where timber harvesting was the main income source. In addition, the concept of sustainable use of public benefit forests is in principle allowed by the forestry departments. However, in practice there is room to interpret whether different activities in public benefit forests correspond with the principle of sustainable use. Governments always try to avoid the burden of monitoring sustainable use of public benefit forests and the risk of farmers' overcutting. As a result, they prefer to simply reject any kinds of utilization of public benefit forests. Controversies often exist between governments and local communities on selective logging and livestock browsing in public benefit forests. The principle of sustainable forest management alone is not enough to reduce negative impacts of the payment schemes on local livelihood. Related policies and institutions have to be established to materialize this principle. This leads to the discussion on another category of factors impacting local livelihood – institutional factors.

Through the investigation of the cases and the comparison between them, institutional factors stand out as a significant force to differentiate the effect of payment schemes on local livelihood. Especially collective forest tenure reform and forestry industry policy played important roles in reshaping the structure of the rural economy, securing payment distribution, and facilitating sustainable use of forest resources.

Collective forest tenure reform reassured the rights of forest owners to obtain payment for giving up their forests for public benefit purpose. Formal responsibility contracts for public benefit management have been signed between the governments and local farmer households. Local farmers had access to participate in designing and deciding the arrangements for sharing the payment between villages and farmer households. As a result, a larger share of the payment was distributed to local farmers than before. It is one of main reasons for the relatively low negative income impacts in Fujian. Furthermore, the tenure reform created favorable conditions for forestry development in forested areas, which boosted forest-related income in the rural economy. To some degree it ameliorated negative impacts of logging limitation on local farmers' income. Another important institutional factor is forestry industry policy, which the provincial forestry departments designed to direct forest resources use and forestry production. If this policy contains measures to harmonize utilization of forest resources and ecological conservation, forestry industry policy can reduce negative impacts of the payment schemes. This has been exemplified in Liaoning's case, where local farmers were encouraged and financially supported to develop agro-forestry in public benefit forest. It has to be noted that stable and definite forest ownership is a precondition for a well-functioning forestry industry policy. The access to public benefit forest resources guarantees a positive incentive to protect forests for agro-forestry utilization.

Finally, contextual factors are also important for determining the scale and degree of the economic impacts of the payment schemes. Although most of the local farmers impacted by the schemes lived in forested areas, their dependence on forest resources varies from region to region. In Fujian, most of the forests demarcated as public benefit forests are natural forests, which have a relatively low economic value and have not been developed by local villages. The communities were less dependent on the public benefit forest for living. In contrast, local farmers in Guangxi were relatively highly dependent on forest resources. They had a long history of extracting timber from forests, and transform natural forests into plantations or bamboo forests with high economic value. Forestry was their main income source and the local economy largely relied on forests. Apart from the dependence on forests, alternative livelihood opportunities such as developing agro-forestry and ecotourism, and off-farm work are also crucial to adjust rural income structure and to compensate the loss due to logging limitations. Alternative economic opportunities are not only endowed by geographical and natural resources conditions but also relate to institutional support or restrictions.

## 7.2.4 Participation

Wunder (2005) defined PES as a voluntary transaction where ecological services are bought by buyers from providers. However, China's PES differs from this standard definition in its nature of compulsory implementation. Although the central government stipulated that the payment schemes should be implemented with full agreement of local communities and farmers, the provincial governments have to some extent unilaterally changed collective and private forests into public benefit forests. There was indeed a consultation process while the governments were designating public benefit forests, but the governments had the final say in deciding the outcome of the demarcation. Therefore, it is crucial to analyze the participation mechanism in China's PES to find out how and to what extent it differs from PES in other countries. Studying participation is also important to explore how legitimacy is established during the implementation of the payment schemes. This study focused on the dynamics and forms of participation, which are introduced, applied, maintained and transformed by the payment schemes in its various stages of formulation, implementation and examination.

This study evaluated how and the degree to which local farmers have been involved in the implementation of the payment schemes. The implementation of payment schemes generally includes four stages: policy formulation, demarcation of public benefit forest, signing forest management contracts, and examination (Table 7.8). The investigation showed that there was no opportunity for farmers to participate in the *formulation* of payment schemes at either the national or the regional level. The payment schemes have been formulated to a large degree within the administrative system, where forestry authorities negotiated with other governmental agencies and municipal governments to facilitate the establishment of payment schemes. In the policy

*Table 7.8. Participation of payment schemes in the three cases.*

| Stages in policy process | | Fujian | Guangxi | Liaoning |
|---|---|---|---|---|
| Formulation | | no | no | no |
| Demarcation | | 94% | 11% | 61% |
| Forest management contract | type 1 government & villages | 81% | 33% | –[2] |
| | type 2 villages & foresters | 96% | –[1] | 69% |
| | type 3 villages & farmers | 90% | 39% | 33%[3] |
| Examination | | 94% | 6% | 33% |

[1] In the Guangxi case, the responsibility for the public benefit forest was with individual farmer households. No professional foresters were hired.

[2] In the Liaoning case, the county government directly signed public benefit forest management contracts with farmer households after the collective forest tenure reform. There was no contract between the government and villages.

[3] This relates to the number of farmer households which have signed contracts with the county forestry bureaus until 2010. The contract signing was still going on during the tenure reform.

formulation of payment schemes, the representatives of Provincial People's Congress and Political Consultation Conference also played a crucial role in putting payment schemes on the political agenda of the provincial governments. However, there was no direct channel for local farmers to participate in policy making for payment schemes.

In the *demarcation* of public benefit forests, local farmers have been consulted about the demarcation result, especially in Fujian, which created trust among local farmers affected by the result and thereby increased the legitimacy of the policy. This process is decisive for the potential impacts of the payment schemes on local livelihood. The demarcation process determines which plots should be selected as public benefit forests and how many forest plots should be included into the schemes. Local participation can increase the appropriateness of demarcation, which means keeping ecologically important forest plots for public benefit purpose and giving economically valuable forest plots to local communities and farmers. The investigation showed a high consultation rate for the Fujian case, where better environmental performance was also achieved. In contrast, a relatively low rate of consultations in the Guangxi case correlated with its low environmental effectiveness. It should be noted, though, that the low environmental effectiveness also related to many other factors such as negative forestry industry policy, conflicts with local traditional livelihoods and recent natural disasters. The demarcation of public benefit forest is a process of redistributing wealth (both economic and environmental) generated from forest ecosystems. The factual acceptance of the demarcation result rests not only with trust and legitimacy by local farmers, but also with the fear of, and submission to, indirectly threats of sanctions from the government. In addition, compliance also could be engendered by the farmers' perception of their own powerlessness and the lack of alternatives open to them. However, in Guangxi, negative forestry industry policy and severe conflicts with local traditional livelihoods induced doubts of local farmers on the demarcation results and created or maintained alternatives diverting from public benefit forest management. Therefore, the lack of consultation and other negative factors worked together to reduce the legitimacy of the demarcation result and contributed to lower environmental effectiveness.

Public benefit forest *management* is the most important component of the payment schemes. In practice, forest management arrangements differ from region to region. In the Fujian case, the government, villages, local farmers and foresters shared the responsibility of forest management together. Three types of contracts have been signed between the government and villages, villages with foresters, and villages with farmers. This arrangement has the advantage in specifying and clarifying different responsibilities for various stakeholders. The responsibility of monitoring and examination rests on the government and villages; the responsibility of professional forest management is undertaken by foresters; local farmers bear the responsibility of stopping timber harvesting and protecting their own forests. In the Guangxi case, local farmers bear all the responsibility of managing and protecting public benefit forests, while villages and governments undertake monitoring and examination. It is doubtful that local farmers can carry out effective management when the payment is so low and much effort is required for taking care of forests (stopping illegal logging, and controlling fire and pest). The evaluation on environmental performance showed the management arrangement in the Guangxi case did not achieve satisfying environmental effectiveness. The Liaoning case relied more on the forestry administrative apparatus for implementing forest management and protection. The forestry bureaus took the responsibility of selecting, hiring, and evaluating foresters. Compared with the other cases, villages cooperated with forestry authorities, with a less important role in organizing foresters and monitoring forest protection practices. In order to implement these management arrangements, the forestry authorities relied on cooperation and participation of local farmers. The majority of local farmers has been consulted for the selection and evaluation of foresters.

In the *examination* for public benefit management, the Fujian case had a very high rate of consultation of local farmers. In the Liaoning case, one third of local farmers have been asked for information and advice on forest management. In Liaoning and Fujian, the provincial and county forestry authorities developed specified procedures to consult local communities and farmers. Formal rules and procedures facilitated participation of local people in the examination of the schemes and thereby improved the environmental performance. In contrast, few local farmers had participated in the examination of the schemes in the Guangxi case, where local participation had not been required as a formal procedure for the examination of the schemes.

Overall, in the Fujian case, a significantly high proportion of local farmers has been consulted during the implementation of the payment schemes including demarcation, forest management contracting and examination. Also in Liaoning a relatively high proportion of local farmers had been consulted, especially in the process of demarcating public benefit forest and making forest management contracts. In the Guangxi case, in contrast, few local farmers were involved in the implementation of the payment schemes. Correspondingly, wide participation generally resulted in high satisfaction with the implementation of the schemes. Table 7.9 shows that in the Fujian case the highest percentage of local farmers was satisfied with the implementation of the schemes, including the demarcation, forest management and examination. Especially in Guanxi, few farmers were satisfied with the implementation of the payment policy.

The principle of participation is one of the important factors influencing the legitimacy – and in the end success – of the payment schemes. In practice, participation is realized by ensuring involvement and consultation of local farmers. The development of forms of democracy in Chinese rural communities is a gradual progress, and provides an institutional setting for the implementation of the payment schemes. Like other programs in rural areas, these schemes follow the principle of representation required by the Organic Law of Village Committees. The

*Table 7.9. Percentage of farmers satisfied with the payment policy process.*

| Cases | Demarcation[1] | Forest management[2] | Examination[3] |
|---|---|---|---|
| Fujian | 88% | 100% | 78% |
| Guangxi | 0% | 40% | 6% |
| Liaoning | 54% | 85% | 69% |

[1] Percentage of respondents who agreed with the result of demarcation.
[2] Percentage of respondents who were satisfied with the arrangement of public benefit forest management (public benefit forests managed by either foresters or farmer households themselves).
[3] Percentage of respondents who were satisfied with the existing examination body – township forestry stations.

law stipulates the basic rules and procedures for representation of village members in collective affairs, including public benefit forest management. However, in practice the degree of participation of local farmers depended on whether formal rules and procedures specified for the payment schemes have been established. More importantly, participation of local farmers improved the appropriateness of selecting forest plots for public benefit forest, avoids severely negative impacts on local livelihood and increased acceptance and compliance with the payment schemes.

### 7.2.5 The specific character of Chinese PES

According to Wunder (2005: 3), PES is "a voluntary transaction where a well-defined ES is being 'bought' by a ES buyer from a ES provider, if and only if the ES provider secures ES provision." However, Chinese PES in forestry is hard to fit in this definition according to their design characteristics and implementation. First, PES in China is not based on voluntary transactions. PES in China is designed and implemented by the government with limited consultation with (and agreement of) local stakeholders. Local farmers, who often have limited access to policy formation and are under political pressure of local authorities, hardly have powerful positions in negotiations with governmental agencies on PES. As a result, local farmers are at best consulted during the establishment of PESs and usually feel governmental pressure to participate in the PES scheme, rather than that they are motivated by the payment to voluntarily participate in the PES scheme, as is the dominant perspective in other countries. Second, farmers have limited land-use choice under the existing land-use policies and forest resources management policies. China's land-use planning does not allow farmers to change land-use. Chinese Forest Law also stipulates that farmers have to obtain forest logging permission from local forestry authorities before starting timber harvesting. Under these conditions, PES can be seen as "minor welfare subsidies to ease the implementation of command-and-control measures" (Wunder et al., 2005: ix). This type of PES is different from PES schemes that use payments as economic incentive to induce positive environmental consequences. Of course, there is a continuum between minor complements to command-and-control measures and pure PES as defined above. NFPP can be regarded as a PES program, which mainly relies on command-and-control measures to change forest use practices and provides subsidies for implementing forest management. CCFGP, on the other side is more close to voluntary transactions, and relies more on economic incentives to induce land-use change. The Public Benefit Forest Compensation Fund is in the middle, as it employs both incentive and regulatory measures to protect public benefit forest and provide ecological services. These characteristics make Chinese PES different from PES in other countries.

In China, and beyond China, forest conservation is frequently seen as a collective good and the responsibility of governments. However, governments do not always know the importance of forest ecological services for beneficiaries. Furthermore, governmental budget can only provide a minimum payment for subsidizing local forest protection. Incomplete information on ecological services and insufficient funding (e.g. less than what beneficiaries would be willing to pay) compromises efficiency of PES. Hence, it is commonly believed that user-financed PES programs are much more likely to be efficient than government-financed ones (Wunder et al., 2008). Although government-financed PES dominates in China's forest sector, some small scale user-financed PES schemes exist, such as conservation contracts between local forest owners and hydropower stations. It is necessary to examine this type of PES and compare it with the government-financed PES that were predominant in our cases, to offer insights in providing alternative for these government-financed PES.

### 7.2.6 Conclusion

Payment schemes have achieved different levels of environmental effectiveness in different locations. According to the comparison of the three cases, environmental effectiveness is highly dependent on the institutional setting, the attributes of the payment schemes and the contextual factors. In the Fujian case, the payment schemes have accomplished a higher environmental effectiveness than those in the other two cases. The better environmental performance benefited from a relatively inclusive demarcation process of public benefit forest, in which forest owners have been widely consulted on the result, and where less forestland that is important for local livelihood has been given up for public benefit forest.

Under the existing payment standard, the schemes tend to have a negative impact on local livelihood. In all three cases, farmer households complained that the payment schemes had a negative impact on their income. Alternative livelihood opportunities, including off-farm employment, agro-forestry, and ecotourism, can partly offset these direct income impacts. At the same time, local forest use practices have shifted away from traditional timber extraction and transformed into a more sustainable use of forest resources. Forest industry policy played an important role in facilitating or reversing this process.

Local forest owners have not been involved in the decision making on payment schemes; neither was the participation of local farmer households in demarcation prevalent. Local farmers still lack influence in shaping the payment schemes and in initiating public benefit forests (e.g. what kind of forest should be included as a public benefit forest; how much is the payment for it). In contrast, farmer households have participated or been included in the management of public benefit forests. There is indication that their involvement in the implementation increased the environmental effectiveness of the payment schemes.

Collective forest tenure reform seems to have improved the environmental performance of payment schemes and have reduced their negative impacts on local livelihood. Distribution of public benefit forests to individual farmer households facilitated the flow of payments to individual farmer households and reduced the risk that the government and village committees divert the payment. Furthermore, the forest tenure reform created new room for developing market-based voluntary PES in China. Clarification of forest ownership provided a helpful institutional setting for establishing a more market-oriented PES mechanism.

## 7.3 Reflection on research and conceptual framework

### 7.3.1 Reflection on the evaluative framework

Although the payment schemes for forest ecological services have been widely applied in China, their performance has not been carefully examined yet. Indeed, there is a lot of literature on PES in China, but most of this literature is either devoted to a general introduction on different kinds of payment schemes or focuses on how to design "reasonable" payment schemes from a purely theoretical perspective (Li *et al.*, 2006b,c; Liu *et al.*, 2007; Zheng and Zhang, 2006). Most of this research concludes that low payment standards is the main problem in the functioning of current payment schemes and recommendations thus point to the development of a more scientific payment standard or a market-based payment mechanism (Chen and Wei, 2007; Li *et al.*, 2007; Zhang *et al.*, 2010a). Few studies have related these analyses and recommendations to empirical examination, with the exception of a few empirical studies based on single cases, such as those of Li and Chen (Li and Chen, 2007) on Hainan, and Chen (Chen, 2006) on Fujian. But even these case studies are limited in only focusing on the evaluation of socio-economic impacts of payment schemes on local people, without paying attention to for instance participation of these local people in the design and implementation of payment schemes and the effect of that on socio-economic and environmental variables. In addition, these single case studies usually pay attention to the payment schemes themselves, neglecting the role of other institutional factors in the functioning of the schemes, such as forest tenure reform, forest management institution, and forestry industrial development policies.

Addressing these problems, this thesis attempted to provide an integrated evaluation of payment schemes based on comparative analyses of multiple cases. In doing so, an evaluative framework was developed to explore the relationship between institutional factors, the design and functioning of payment schemes, their environmental and economic performance and the role that participation played in that. The basic reasoning behind this research framework was that payment schemes are introduced in an existing (but often dynamic) institutional setting and this combination of payment scheme introduction cum institutional setting changes local forest use practices but also the institutional setting. This process sets up a set of rules on local participation and has environmental consequences and economic implications for local forest owners. And these relations and effects are better understood in comparing different cases.

This framework brought a number of advantages to this research. First, separating an institutional setting from the payment schemes provides conceptual clarity to analyze the interaction between the payment schemes and other relevant institutions. This is especially helpful since the institutional setting in China is unstable due to – in our empirical field – a series of reforms in the forest sector. Since this research started in 2007, a collective forest tenure reform has been initiated in several provinces and it brought significant changes to the structure of forest property rights in rural areas. Moreover, the payment schemes themselves also have been adjusted on payment standard and distribution due the research period. Our multiple case research design made it possible to compare different (changes in) institutional settings and to explore relations and causalities between change in forest use practice and institutional factors. Second, the research framework did introduce forest use practices into the evaluation, which focuses on rather directly measuring the impacts of payment schemes on the environment and local livelihood. This deviates from previous research which focuses very much on output indicators in terms of environment and household economics. Forest use practices, as an inclusive concept, helps to conceptualize the change in the practices and modes of forest production and consumption induced by the payment schemes. Analyzing this change in the mode of use practices provides a lot of details of the process of ecological modernization through payment schemes in a specific local context (how does the forest use change from timber harvesting to developing eco-tourism, agro-forestry and other sustainable uses, how are daily management systems on public benefit forest set up and functioning, how are payments divided among the different actors and organizations, etc.). This is largely neglected by those earlier researches mentioned above, which only emphasize performances and consequences. Third, the framework includes participation of farmers as a procedural and an evaluative aspect. Participation is directly linked to the institutional setting and affects the design of PES, its functioning and its effectiveness (Beckmann *et al.*, 2009). Using perspectives of political modernization as included in Ecological Modernization Theory, the study assumes that the involvement of local farmers in the implementation of payment schemes can reduce negative impacts on farmers' livelihood, increase the legitimacy of the introduced PES schemes and as such improve the effectiveness of PES. This inclusion of participation and mechanism and outcome fills in the theoretical and empirical gap in the existing literature on China's PES, for forests and beyond. The nature of participation is one of the things that makes China's PES different from PES in other countries. Moreover, our focus on participation in PES contributes also to wider studies on participation and democracy at the local level in rural China.

It is worth noticing also some challenges in implementing the research and shortcomings in the evaluative framework. First, the research had to face an unstable institutional setting. China's forest sector is experiencing rapid institutional transformations, including collective forest tenure reform, logging management reform, and policy adjustment in ecological conservation projects. These transformations caused a series of discontinuities in the institutional setting of forest management. Collective forest reform changed the property rights structure of collective forest and logging management reform is turning the old logging quota management system into a forest management plan system. Furthermore, other ecological conservation programs such as CCFGP and NFPP also reshaped the payment policy domain by being integrated into the payment schemes, and also competing for financial resources with them to provide policy alternatives. It proved complex to clarify how, to what extent and with what "effects" these institutional changes interacted with each other and with the implementation of the payment schemes.

Second, this research is based on a kind of rational institutional theory, which takes institutional factors (rules, incentives and power) as drivers for changes of forest use practices. It indeed managed to contribute to explanations of how forest use has changed and why such change happened in specific local contexts. The payment schemes not only set up a series of rules to manage forests in a more sustainable way, but also provide financial resources to encourage such transformation in forest management. The framework also helped in noticing the role of power in China's forest administration system, which helped in explaining the functioning of the payment schemes the rather low payments compared to the expectations of local forest owners. While these institutional factors have been examined thoroughly in this research, the analytical framework ignored a cultural perspective on the traditions, norms, ideas and discourse about forest management and use and how these rather stable cultural dimensions (or the first dimension of Williamson's (2000) institutional structure) influence payment schemes in local contexts. Some research has demonstrated the role of traditions and norms in community forest management (Xu and Melick, 2007). Adding a cultural dimension to the conceptual framework could provide new insights on how traditions and norms of play a role in reshaping forest management and use during the transformation of ecological modernization in China's forest sector. It could perhaps also add insight in the "fitting" of PES in (different) Chinese circumstances.

### 7.3.2 Reflection on research methods

Compared to previous literature, which either provides a general discussion on PES mechanisms based on a theoretical perspective or examines a single case, this study utilizes multiple cases and empirical data obtained via famer household surveys and interviews with key stakeholders in different regions. The trans-regional dataset offers insights into different institutional settings, including different forest tenure structures after the tenure reform and various localized payment scheme designs. To guide the analysis, a conceptual framework has been developed to address three aspects of payment schemes (environmental effectiveness, economic and livelihood impacts, and participation). The operationalization of these evaluative factors had to take into account the existing availability of forest resource and environmental data, which were usually inconsistent or even not record. To repair these data quality and availability problems, the research has to use local farmer observations and subjective assessments as indicators for environmental quality and changes in environmental qualities. Data from county or provincial authorities were also collected to carry out a basic "verification" whether the assessment of local farmers on case sites parallels with general developments and changes in the whole region. While, given the chosen research design and the availability of time and resources, it is the best what could be done in terms of assessing environmental effectiveness of the schemes, it is far from ideal. Better and more forest resources and environmental data would enable stronger conclusions as to what have been the environmental consequences of the PES schemes.

The availability of more time and resources would also have enabled the inclusion of more interviewed farmers per case study and the inclusion of more case study sites per province. The former would have given a better internal validity and would have enabled better (statistical) analyses of the environmental and livelihood effects. The latter would have contributed to furthering the external validity and the representativeness of our cases for the provinces included in this study. The inclusion of more provinces would even have enabled me to draw China wide conclusions. But more case studies could also have enabled us to sort out the contribution of various institutional and contextual factors on the (environmental and economic) performances of the schemes.

Regardless of this "wish list" of the quality and quantity of data on local payment schemes, the case studies carried out in this research do have wider relevance for understanding payments schemes in the Chinese forest sector. Other provinces in China also have the same forestry administrative structure and procedures for implementing payment schemes and they shared a similar history of developing payment schemes as the selected provinces. In addition, the forest tenure reform is also carried out nationwide. As analyzed above, the dynamics of payment schemes are determined by these factors. Hence, one can to some extent expect similar results on the performance of payment schemes in other provinces.

### 7.3.3 Reflection on Ecological Modernization Theory for China's forest sector

The evaluation showed that the payment schemes in China can be understood as a process of ecological modernization in China's forest sector. Forest resources have been protected from the threat of clearing and unsustainable use. Forest ecosystems have been maintained for generating ecological services and products, rather than only for producing timber. Economic incentives have been employed as an important instrument to encourage the change of forest use practices from timber harvesting into protection for ecological functions. And a variety of actors, beyond the state, have been included in these sustainable forest management arrangements. All developments have been indicated by EMT under the notions of economizing ecology and political modernization, although they are colored in the case of payments schemes with typical Chinese characteristics (e.g. Zhang *et al.*, 2007). Compared to ecological modernization inspired PES schemes in western countries, regulatory measures are still playing an important role in maintaining the Chinese schemes and reigning local forest use. The payments alone cannot guarantee a smooth transformation of forest use practices. The government plays a major role in steering the direction of ecological modernization processes through its administrative system and its financial support. Market actors (e.g. farmers, private forest industries, tourist industry) play less significant roles in the process than in the PES schemes of other countries. However, the trend in Chinese forest policy seems to be that the strength of economic incentives and payment schemes will increase gradually, while regulatory measures might become less dominant and decisive in the future.

As mentioned in Chapter 2, there are doubts from neo-Marxist scholars on such ecological modernization inspired schemes, as marginal local farmers might become further marginalized during processes of ecological modernization in – in our case – the forest sector. This research showed that it is not evident that the implementation of payment schemes automatically traps local farmers into more disadvantaged positions, although some local farmers indeed endured income loss during the implementation. It depends very much on how payment schemes have been designed and implemented, to what degree local farmers can been involved in the implementation, and how other institutional factors such as forest tenure reform and forest industry policy interact with the payment schemes. The payment schemes could as well be a process of empowerment to local farmers; and the collective forest tenure reform seems to have enhanced such an empowerment.

Through a systematic examination of environmental effectiveness, livelihood impacts and participation mechanism of Chinese PES, the research confirms that an ecological modernization perspective can be used fruitfully for understanding and explaining the current ecological transition in the Chinese forest sector, of which the introduction of payments are part. It would be fruitful to further compare this process of ecological modernization in China's forestry sector with similar processes in other sector, in further characterizing and contributing to a "Chinese style" Ecological Modernization Theory.

## 7.4 Implication and recommendations for payment policies

What has become clear from this thesis is that payment schemes have developed into important policy arrangements for sustainable forest management in China. The ability to develop and implement payment schemes is set by national policies, but the specific arrangements and design of the schemes are co-determined by local policies and the specific characteristics of the institutional setting. From our case studies several recommendations can be formulated for further developing PES in Chinese forest management.

Regulation on participation should be formulated to improve and guarantee the involvement of local forest owners in the design and implementation of payment schemes. Participation in payment schemes has been imprinted with Chinese characteristics – government directs the formulation and design of policy, local stakeholders are at best consulted, and generally the government has the power to decide finally. Although the central authority – the State Forestry Administration – suggested that the demarcation of public benefit forest should be done with permission of local farmers, this "principle" has not been uniformly applied in practice. In some regions, provincial governments just demarcated public benefit forest by their own technical staff without consultation or agreement of local forest owners and farmers. In addition, local farmers usually lack information on payment policies and capacity to bargain with the government on final decisions. Hence, in order to improve that a formal and practical rule of participation should be integrated into the payment schemes. This can effectively increase the legitimacy of the payment schemes, reduce negative impacts on local livelihood and improve their environmental performance in the end. In addition, the ongoing forest tenure reform is creating new small forest owners. Public benefit forest has been fragmented and more and more small stakeholders have been involved in the policy domain. A formal rule of participation also helps to smoothen the transfer of ownership on public benefit forests and establish formal management and protection contracts with local forest owners.

The payment schemes at national and provincial level already changed local forest practices from harvesting timber to providing ecological services. However, the payment is determined by government rather than by market-based instruments. Although the Forest Law encourages the application of market-based mechanisms for pricing forest ecosystem services, there are still institutional and informational barriers, such as ambiguous forest ownership, weak willingness of beneficiaries to pay, and unclearly defined ecological services. Some barriers are fading away following institutional changes in the forest sector such as the forest tenure reform, while others should be solved with capacity building of local forest authorities in the long run. More market-based mechanism can be expected to increase the efficiency of the payment schemes, which now only use a single standard for the payment. In the process of institutionalization of the payment schemes, space should be reserved for developing and including such market-based mechanisms. For example, auction and bidding mechanisms can be introduced for highly valuable public benefit forests and through such mechanisms additional funding for protection can be provided. Alternatively, within the range of public benefit forests, forest authorities can allow and encourage environmental NGOs or other donors to make conservation contracts with local forest owners for more diverse ecological services such as carbon sequestration and biodiversity, instead of only relying on governmental funding for forest protection.

Cross-sector and cross-boundary cooperation (between forestry, water conservation, and environmental protection departments, for instance) should be established to facilitate an integrated payment scheme for ecosystem services, instead of the existing single sector payment models. China's environmental protection domain has been divided into different sectors and each sectoral administration separately takes responsibilities to deal with environmental issues in its own sector. With respect to PES, forestry, water conservation, and environmental protection departments have all established payment schemes for ecosystem services. However, there is neither substantial cooperation between these payment schemes, nor is an integrated payment scheme designed to involve different sectors. The division of sectors and ineffective cooperation between different regions also causes the separation of providers and beneficiaries of environmental services and makes it more difficult to reach PES agreements between them. For example, forestry departments usually have controversies with water conservation departments, which are also in charge of hydropower station and reservoirs, on the contribution of these beneficiaries. Western regions, which are usually providers of ecological services, and less developed areas often blame eastern and richer regions of avoiding their responsibility of providing payment for ecological services. Currently (2011) the central government is drafting a regulation on eco-compensation, which emphasizes setting up rules for cross-sector and cross-boundary cooperation on PES. It is still unclear how this rule can be implemented within the existing institutional setting in such a way that it breaks barriers between different sectors and/or regions.

## 7.5 Recommendations for further research

This contribution has been based only on a small sample of schemes and counties and on a case study methodology. In addition, also within the cases limited numbers of farmer households have been included in the samples. Its findings show the difference in the performance and implementation of payment schemes in different cases and also among different farmer households. The comparison among cases demonstrates the important role of institutional factors, the different components of payment schemes, and the local context on the performance. These findings need to be further investigated and validated by a larger number of samples (both in terms of farmer households, counties and provinces, and payment schemes) in order to generalize the findings for China. This will then also include different research methodologies, more of a quantitative nature.

The structural change in China's forest sector creates new opportunities for new actors engaging in payment policy making and implementation. Climate change, as a discourse, provides an opportunity to restructure China's payment scheme into a new direction. This change could include the emergence of new marketed-based dynamics in payment schemes in the forest policy domain, which has been dominated by governmental directed schemes hitherto. Therefore, future research should focus on how to design and implement the change from governmental directed payment schemes to more marketed-based payment schemes.

Due to limits on time and efforts, this study only focused on payment schemes within the forest policy domain. However, other policy domains also influence these payment schemes or have developed other payment policies that also relate to forest ecosystem protection, such as watershed compensation schemes initiated by environmental protection department. The specialization, professionalization and fragmentation of governmental bodies causes a lot of problems in integrating these payment schemes. Further research is needed to explore in detail how such interdependencies now work and how this situation can be improved.

# References

Adam, S. and Kriesi, H., 2007. The network approach. In: P. Sabatier (ed.), *Theories of the policy process*. Boulder, CO, USA: Westview Press, pp. 129-154.

Arts, B. and Buizer, M., 2009. Forests, discourses, institutions: a discursive-institutional analysis of global forest governance. *Forest Policy and Economics*, 11 (5-6), 340-347.

Arts, B. and Van de Graaf, R., 2009. *Theories for forest policy: an overview*. In: W. De Jong (ed.) Forest policies for a sustainable humanosphere. Kyoto, Japan: Center for Integrated Area Studies, Kyoto University, pp. 35-40.

Beck, U., 1992. *Risk society ; towards a new modernity*. London, UK: Sage Publications, 260 pp.

Beckmann, V., Eggers, J. and Mettepenningen, E., 2009. Deciding how to decide on agri-environmental schemes: the political economy of subsidiarity, decentralisation and participation in the European Union. *Journal of Environmental Planning and Management*, 52 (5), 689-716.

Bennett, M.T., 2008. China's sloping land conversion program: Institutional innovation or business as usual? *Ecological Economics*, 65 (4), 699-711.

Benxi County Government, 2005. *Benxi County Public Benefit Forest Management Measures*. People's Government of Benxi Man Autonomous County.

Blamey, A. and Mackenzie, M., 2007. Theories of change and realistic evaluation. *Evaluation*, 13 (4), 439-455.

Campbell, D. and Stanley, J., 1963. *Experimental and quasi-experimental designs for research*. Chicago, IL, USA: Rand McNally.

Central Committee of Chinese Communist Party and Chinese State Council, 1981. *Decision on several issues on forest protection and forestry development*. The Central Committee of Chinese Communist Party and the State Council.

Chen, G., 2000. Contemporary elite power and power structure in China's rural communities. *China 21*, 8 (5), 163-190.

Chen, Q., 2006. *Research on eco-compensation of public benefit forest*. Beijing, China: China Forestry Publishing House.

Chen, Q. and Wei, Y., 2007. Standard, scope and method of ecological compensation related with non- commercial forest. *Science and Technology Review*, 25 (10), 64-66.

Chia, E., 2009. *State policy intervention and perceptions of environmental (and civic) responsibility in forest policing: case study of a township in China*. Australian Political Studies Association Annual Conference 2009. Sydney, Australia.

China Center for Modernization Research, 2007. *China Modernization Report 2007: study of ecological modernization*. Beijing, China: Peking University Press.

Chinese State Council, 1992. *Notice on important points of economic system reform in 1992*. the State Council.

Chinese State Council, 2001. *Regulation on the Implementation of the Forest Law*. the State Council.

Chinese State Council, 2003a. *Decisions regarding speeding up forestry development*. the State Council.

Chinese State Council, 2003b. *Regulation on CCFGP*. the State Council.

Chinese State Council, 2005. *Decision on implementing scientific concept of development and strengthening environmental protection*. the State Council.

Connora, J.D., Warda, J., Cliftonb, C., Proctora, W. and MacDonald, D.H., 2008. Designing, testing and implementing a trial dryland salinity credit trade scheme. *Ecological Economics*, 67 (4), 574-588.

Crabbé, A. and Leroy, P., 2008. *The handbook of environmental policy evaluation*. London, UK: Earthscan.

Dai, L., Zhao, F., Shao, G., Zhou, L. and Tang, L., 2009. China's classification-based forest management: procedures, problems, and prospects. *Environmental Management*, 43 (6), 1162-1173.

Dai, S.B., Yang, S.L. and Cai, A.M., 2008. Impacts of dams on the sediment flux of the Pearl River, southern China. *CATENA*, 76 (1), 36-43.

DECC, 2007. *Biodiversity banking and offsets scheme: scheme overview*. Sydney, Australia: Department of Environment and Climate Change.

Economy&Nation Weekly, 2010. *Hot debate between the government and the academia: is* Eucalyptus *the cause of the drought?* Available at: http://finance.sina.com.cn/g/20100412/14377733126.shtml.

EEA, 2001. *Reporting on environmental measures: are we being effective?* Luxembourg, Luxembourg: European Environment Agency.

Environmental politics, 2006. *Special issue on environmental governance in China*, 15. Abingdon, UK: Routledge.

EUROPA, 2008. *Support for rural development by the European Agricultural Fund for Rural Development (EAFRD)*. Available at: http://europa.eu/scadplus/leg/en/lvb/l60032.htm.

FAO, 2010. *Global Forest Resources Assessment 2010*. Rome, Italy: Food and Agricultural Organization of the United Nations.

FAO, 2011. *State of the World's Forests 2011*, Rome, Italy: the Food and Agriculture Organization of the United Nations.

Forest History Society, 2006. *The Weeks Law*. Available at: http://www.foresthistory.org/Research/usfscoll/policy/Agency_Organization/NF_System/weeks_law/index.html.

Forestry Commission, 2008. *English woodland grant scheme*. Available at: http://www.forestry.gov.uk/ewgs.

Fujian Provincial Forestry Department, 2005. *Planning outline of Fujian provincial forest for ecological public benefit*. Fujian Provincial Forestry Department.

Fujian Provincial Government, 2005. *Fujian province's management measure on public ecological benefit forest*. Fujian Provincial Government.

Fujian Provincial Government, 2007. *Opinion on promoting management mechanism reform for public benefit forest*: Fujian Provincial Government.

Gao, Q., 2006. Discussion on soil and water conservation of Moso bamboo. *Subtropical Soil and Water Conservation*, 18 (2), 39-40.

Gobbi, J.A., 2000. Is biodiversity-friendly coffee financially viable? An analysis of five different coffee production systems in western El Salvador. *Ecological Economics*, 33 (2), 267-281.

Greenpeace, 2010. *What makes drought worse in Guangxi? The secret of fast growth of* Eucalyptus *forest* Available at: http://www.greenpeace.org/china/zh/news/gx-plantation-story.

Grieg-Gran, M., Porras, I. and Wunder, S., 2005. How can market mechanisms for forest environmental services help the poor? Preliminary lessons from Latin America. *World Development*, 33 (9), 1511-1527.

Guangxi Forest Tenure Reform Office, 2009. *Notice on improving current forest tenure reform in Guangxi*. Guangxi Forest Tenure Reform Office.

Guangxi Public Benefit Forest Protection and Development Task Force, 2009. *Investigation report on Guangxi public benefit forest protection and development*. Guangxi Public Benefit Forest Protection and Development Task Force.

Guangxi Regional Government, 2008. *Ecological function zoning plan of Guangxi Zhuang autonomous region*. Guangxi Regional Government.

Guttman, D. and Song, Y., 2007. Making central-local relations work: comparing America and China environmental governance systems. *Frontiers of Environmental Science and Engineering in China*, 1 (4), 418-433.

GW and EIA, 2010. *Investigation into the global trade in Malagasy Precious woods: rosewood, ebony and pallisander*. Global Witness and the Environmental Investigation Agency (US).

Hamilton, K., Bayon, R., Turner, G. and Higgins, D., 2007. *State of the voluntary carbon market 2007: picking up steam*. Washington, DC, USA: New Carbon Finance and The Ecosystem Marketplace.

Han, J., 2009. *Renewable energy development in China: policies, practices and performance*. Wageningen, the Netherlands: Wageningen University.

Hao, Y. and Ou, G., 2008. *Investigation report on damage to Guangxi's forestry by snow disaster in 2008.* Beijing, China: State Forestry Administration.

Hardin, G., 1968. The tragedy of the commons. The population problem has no technical solution; it requires a fundamental extension in morality. *Science*, 162 (859), 1243-1248.

He, J.Y. and Wu, R.H., 2001. Theory of ecological modernization and current environmental decision-making of China. *Journal of Chinese Population, Resources and Environment*, 11 (4), 17-20.

Hill, M., 2005. *The public policy process.* Harlow, UK: Pearson Longman.

Hong, D., 2010. Chinese Environmental Sociology: Some Remarks. *Academia Bimestris*, 2010 (2), 57-59.

Huan, Q., 2007. Ecological modernisation: a realistic green road for china? *Environmental Politics*, 16 (4), 683-687.

Huan, Q., 2010. Growth economy and its ecological impacts upon China: an eco-socialist analysis. In: Q. Huan (ed.), *Eco-socialism as politics*. New York, NY, USA: Springer, pp. 191-203.

Huang, J., 2006. *Address on a matching reform for the collective forest tenure reform–pilot on innovating management mechanisms for public benefit forest.* Fujian Provincial Department of Forestry.

Huang, X., 2009. *2009 Fujian provincial government working report.* Available at: http://www.fujian.gov.cn/zwgk/szfld/hxjsz/jhwl/200902/t20090206_109564.htm.

Huang, Y.N. and Ye, P., 2001. A review on the ecological modernization theory in Western Countries. *Foreign Social Science* (4), 1-6.

Huber, J., 1991. Ecological modernisation: beyond scarcity and bureaucracy. In: A.P.J. Mol, D.A. Sonnenfeld and G. Spaargaren (eds.), *Ecological modernisation reader: environmental reform in theory and practice*. London, UK: Routledge, pp. 42-55.

IIED, 2004. *Markets for environmental services case studies.* Available at: http://www.iied.org/SM/eep/projects/mes/mes.html.

Ingram, H., Schneider, A. and DeLeon, P., 2007. Social construction and policy design. In: P. Sabatier (ed.), *Theories of the policy process*. Boulder, CO, USA: Westview Press, pp. 93-126.

Kaimowitz, D., 2004. Forests and water: a policy perspective. *Journal of Forest Research*, 9 (4), 289-291.

Kosoy, N., Corbera, E. and Brown, K., 2008. Participation in payments for ecosystem services: case studies from the Lacandon rainforest, Mexico. *Geoforum*, 39 (6), 2073-2083.

Kumar, R., 2005. *Research methodology: a step-by-step guide for beginners.* Thousand Oaks, CA, USA: Sage Publications, 332 pp.

Landell-Mills, N. and Porras, T.I., 2002. *Silver bullet or fools's gold? A global review of markets for forest environmental services and their impact on the poor.* London, UK: International Institute for Environment and Development.

Lang, G., 2002. Forests, floods, and the environmental state in China. *Organization and Environment*, 15 (2), 109-130.

Lazdinis, M., Carver, A., Carlsson, L., Tonisson, K. and Vilkriste, L., 2004. Forest policy networks in changing political systems: case study of the Baltic states. *Journal of Baltic Studies*, 35 (4), 402-419.

Lei, J., 2007. *China's forest ecosystem management – a road to forestry sustainable development.* Beijing, China: China Forestry Publishing House.

Li, F. and Chen, H., 2007. An analysis on socio-economic impacts of forest eco-compensation mechanism in Hainan Province. *China Population, Resources and Environment*, 17 (6), 113-118.

Li, S., Tu, X. and Zhang, H., 2006a. *China forestry handbook*. Beijing, China: China Forestry Publishing House.

Li, W., 1999. Flood of Yangtze River and ecological restoration. *Journal of Natural Resources*, 14 (1), 1-9.

Li, W., Li, F., Li, S. and Liu, M., 2006b. The status and prospect of forest ecological benefit compensation. *Journal of Natural Resources*, 21 (5), 677-688.

Li, W., Li, F., Li, S. and Liu, M., 2007. Forest eco-compensation mechanisms and policies options. *Ecological Economy (in Chinese)* (11), 151-159.

Li, X., Jin, L. and Zuo, T., 2006c. *Payment for watershed services in China: the role of government and markets, a diagnositc study, Sustainable sloping lands and watershed management conference.* Luang Prabang, Lao PDR.

Li, Z., Wang, H. and Zheng, Y., 2000. Poverty in forest rich regions. *Forestry economics* (4), 1-7.

Liaoning Provincial CCP Committee and Liaoning Provincial Government, 2009. *Opinion on further deepening collective forest tenure reform and speeding up modern forestry development.* Liaoning Provincial Chinese Communist Party Committee and Liaoning Provincial Government.

Liaoning Provincial Forestry Department, 2006a. *Liaoning Province's detailed rules for implementing the management of public ecological benefit forest.* Liaoning Provincial Forestry Department.

Liaoning Provincial Forestry Department, 2006b. *Liaoning Provincial forestry department's report on prior PBF management and examination in 2006.* Liaoning Provincial Forestry Department.

Liaoning Provincial Forestry Department, 2009. *Liaoning Provincial Forestry department's report on prior PBF management and examination in 2009.* Liaoning Provincial Forestry Department.

Liaoning Provincial Forestry Department, 2010. *Report on collective forest tenure reform in Liaoning Province in 2009.* Liaoning Provincial Forestry Department.

Liaoning Provincial Government, 2005a. *Liaoning provincial government's notice on deepening collective forest tenure reform.* Liaoning Provincial Forestry Department.

Liaoning Provincial Government, 2005b. *Notice on speeding up forestry development in the eastern region of Liaoning Province.* Liaoning Provincial Forestry Department.

Liaoning Provincial Government, 2008. *Liaoning Province development plan for doubling forestry and related industry.* Liaoning Provincial Government.

Lingdian Company, 1999. *Investigation on China's public environmental awareness of forest-related ecological issues.* Available at: http://www.yesky.com/busnews/216455356602122240/19990310/1249442.shtml.

Liu, C., Wang, S., Zhang, W. and Liang, D., 2007. Compensation for forest ecological services in China. *Forestry Studies in China*, 9 (1), 68-79.

Liu, D., 2001. Tenure and management of non-state forests in China since 1950: a historical review. *Environmental History*, 6 (2), 239-263.

Liu, D., 2008. Policies and programmes to make decentralization effecitve: a case study from China. In: C.J.P. Colfer, G.R. Dahal and D. Capistrano (eds.), *Lessons from forest decentralization: money, justice and the quest for good governance in Asia-Pacific.* London, UK: Earthscan, pp. 83-100.

Liu, Y., Mol, A. and Chen, J., 2004a. Material flow and ecological restructuring in China. *Journal of Industrial Ecology*, 8 (3), 103-120.

Liu, Y., Mol, A.P.J. and Chen, J., 2004b. Material flow and ecological restructuring in China: the case of phosphorus. *Journal of Industrial Ecology*, 8 (3), 103-120.

Liu, Z., 1999. Discussion on improving project management of state-owned forest farms in Guangxi. *Guangxi Forestry* (6), 15-16.

LTA, 2006. *The 2005 national land trust census report.* Washington, DC, USA: Land Trust Alliance.

Lu, S., Xiang, W., Li, X. and Qu, z., 2002. Hydrological Characteristics and ecological function estimation of upper reaches of Lijiang River. *Bulletin of Soil and Water Conservation*, 22 (5), 24-28.

Lu, Y., 2007. *Deepening reforms and innovating mechanisms for the protection of public benefit forst: address on the provincial working conference for the innovation on mechanisms for public benefit forest management.* Fujian Provincial Forestry Department.

Marsh, D. and Stoker, G., 2002. *Theory and methods in political science*. New York, NY, USA: Palgrave McMillan.

McAfee, K. and Shapiro, E.N., 2010. Payments for ecosystem services in Mexico: nature, neoliberalism, social movements, and the State. *Annals of the Association of American Geographers*, 100 (3), 579-599.

Merenlender, A.M., Huntsinger, L., Guthey, G. and Fairfax, S.K., 2004. Land trusts and conservation easements: who is conserving what for whom? *Conservation Biology*, 18, 65-76.

Miao, G. and West, R., 2004. Chinese collective forestlands: contributions and constraints. *International Forestry Review*, 6 (3-4), 282-298.

Ministry of Finance, 2007. *Management measures for compensation funding for public benefit forest*. Ministry of Finance.

Ministry of Finance and State Forestry Administration, 2004. *Measure on the management of the central forest ecological benefit compensation fund*. Ministry of Finance and State Forestry Administration.

Miranda, M., Porras, I.T. and Moreno, M.L., 2003. *The social impacts of payments for environmental services in Costa Rica. A quantitative field survey and analysis of the Virilla watershed*, London, UK: International Institute for Environment and Development.

Mol, A.P.J., 1995. *The refinement of production: ecological modernization theory and the chemical industry*. Utrecht, the Netherlands: Van Arkel.

Mol, A.P.J., 1999. Ecological modernization and the environmental transition of Europe: between national variations and common denominators. *Journal of Environmental Policy and Planning*, 1 (2), 167-181.

Mol, A.P.J., 2006. Environment and modernity in transitional China: frontiers of ecological modernization. *Development and Change*, 37 (1), 29-56.

Mol, A.P.J., 2009. Environmental governance through information: China and Vietnam. *Singapore Journal of Tropical Geography*, 30 (1), 114-129.

Mol, A.P.J., 2010. Sustainability as global attractor: the greening of the 2008 Beijing Olympics. *Global Networks*, 10 (4), 510-528.

Mol, A.P.J. and Buttel, F.H., 2002. *The environmental state under pressure. Research in social problems and public policy, volume 10*. Bradford, UK: Emerald Group Publishing.

Mol, A.P.J. and Carter, N.T., 2006. China's environmental governance in transition. *Environmental Politics*, 15 (2), 149-170.

Mol, A.P.J. and Jänicke, M., 2009. The origins and theoretical foundations of Ecological Modernisation Theory. In: A.P.J. Mol, D.A. Sonnenfeld and G. Spaargaren (eds.), *Ecological modernisation reader: environmental reform in theory and practice*. New York, NY, USA: Routledge, pp. 17-27.

Mol, A.P.J. and Sonnenfeld, D.A., 2000. *Ecological modernisation around the world: persectives and critical debates*. London, UK: Frank Cass, 300 pp.

Mol, A.P.J. and Spaargaren, G., 2000. Ecological modernisation theory in debate: a review. *Environmental Politics*, 9 (1), 17-49.

Mol, A.P.J. and Spaargaren, G., 2005. From additions and withdrawals to environmental flows: reframing debates in the environmental social sciences. *Organization and Environment*, 18 (1), 91-107.

National Bureau of Statistics, 2010. *China Rural Statistics Yearbook 2010*. Beijing, China: China Statistics Press.

NCDC, 1998. *Flooding in China, Summer 1998*. Available at: http://lwf.ncdc.noaa.gov/oa/reports/chinaflooding/chinaflooding.html#SUMMARY.

NPC, 1998. *Forest Law of People's Republic of China*. Beijing, China: the standing committee of the National People's Congress.

Ostrom, E., Gardner, R. and Walker, J., 1994. *Rules, games, and common-pool resources*. Ann Arbor, MI, USA: University of Michigan Press.

Pagiola, S., Arcenas, A. and Platais, G., 2005. Can payments for environmental services help reduce poverty? An exploration of the issues and the evidence to date from Latin America. *World Development*, 33 (2), 237-253.

Patton, M., 1994. Developmental evaluation. *Evaluation Practice*, 15 (3), 311-319.

Pawson, R. and Tilley, N., 2003. *Realistic evaluation*. London, UK: Sage.

Pereira, C. and Novotny, S., 2010. Payment for environmental services in the Amazon forest: how can conservation and development be reconciled? *Journal of Environment and Development*, 19 (2), 171.

RMW, 2002. *China flood along Yangtze River in 1998*. Available at: http://env.people.com.cn/GB/41909/42116/3472141. html.

Rogers, P. and Weiss, C., 2007. Theory based evaluation: reflections ten years on: theory based evaluation: past, present, and future. *New directions for evaluation*, 2007 (114), 63-81.

Rogers, P., Petrosino, A., Huebner, T. and Hacsi, T., 2000. Program theory evaluation: practice, promise, and problems. *New directions for evaluation*, 2000 (87), 5-13.

Rojas, M. and Aylward, B., 2003. *What are we learning from experiences with markets for environmental services in Costa Rica? A review and critique of the literature*. London, UK: International Institute for Environment and Development.

Sabatier, P., 2007. *Theories of the policy process*. Boulder, CO, USA: Westview Press.

Scherr, S., White, A., Kaimowitz, D. and Trends, F., 2004. *A new agenda for forest conservation and poverty reduction: making forest markets work for low-income producers*. Washington, DC, USA: Forest Trends.

Scherr, S.J., Bennett, M.T., Loughney, M. and Canby, K., 2006. *Developing future ecosystem service payments in China: lessons learned from international experience*. Washington, DC, USA: Forest Trends.

Scriven, M., 1991. *Evaluation thesaurus*. Newbury Park, CA, USA: Sage.

Shapiro, J., 2001. *Mao's war against nature: politics and the environment in revolutionary China*. New York, NY, USA: Cambridge University Press.

Singer, B., 2008. Putting the national back into forest-related policies: the international forests regime and national policies in Brazil and Indonesia. *International Forestry Review*, 10 (3), 523-537.

Spaargaren, G. and Mol, A., 1992. Sociology, environment, and modernity: ecological modernization as a theory of social change. *Society and Natural Resources*, 5 (4), 323-344.

Stark, T. and Cheung, S.P., 2006. *Sharing the blame: global consumption and China's role in ancient forest destruction*. Greenpeace International and Greenpeace China.

State Forestry Administration, 1999. *Forestry development in China from 1949-1999*. Beijing, China: China Forestry Publishing House.

State Forestry Administration, 2006a. *Investigation report on collective forest tenure reform*, Beijing, China: Department of Policy and Regulation, SFA.

State Forestry Administration, 2006b. *A report for monitoring and assessment of the socio-economic impacts of China's key forestry programs*. Beijing, China: China Forestry Publishing House.

State Forestry Administration, 2009a. *China's conversion of cropland into forest – a decade for CCFGP*. Beijing, China: Office for CCFGP, SFA.

State Forestry Administration, 2009b. *China forest resources report–the 7th national forest inventory*. Beijing, China: China Forestry Publishing House.

State Forestry Administration, 2009c. *Forest resources in China – the 7th national forest inventory*. Beijng, China: State Foresty Adminstration.

State Forestry Administration, 2010a. *China forestry development report 2010.* Beijing, China: China Forestry Publishing House.

State Forestry Administration, 2010b. *Implementation plan on NFPP in the second phase for Northeast China and Inner Mongolia.* Department of Planning and Financial Management, SFA

State Forestry Administration, 2010c. *Implementation plan on NFPP in the second phase for the upper reaches of the Yangtze River and the upper-middle reaches of the Yellow River.* Department of Planning and Financial Management, SFA.

State Forestry Administration, 2011. *China forestry statistical yearbook.* Beijing, China: China Forestry Publishing House.

State Forestry Administration and Ministry of Finance, 2004. *Measures on demarcation of key public benefit forest.* State Forestry Administration and Ministry of Finance.

State System Reform Commission and Ministry of Forestry, 1995. *Outline for forestry economic system reform.* State System Reform Commission and Ministry of Forestry.

Sun, C. and Chen, L., 2006. *A study of policies and legislation affecting payments for watershed services in China.* London, UK: Research Center of Ecological and Environmental Economics, Beijing and International Institute for Environment and Development.

Sun, C. and Chen, X., 2002. *A policy analysis of the China forest ecological benefit compensation fund.* In: CCICED Western China Forest Grasslands Task Force (Ed.), *Workshop on Payment Schemes for Environmental Services.* Beijing, China.

Sun, X., Ni, J., Zhang, W., Yu, X. and Geriletu, 2008. Research on sustainable management technology and ecological benefit evaluation of ecological public welfare Moso bamboo stand. *Journal of Bamboo Research,* 27 (1), 32-37.

Swallow, B., Leimona, B., Yatich, T., Verlarde, S.J., and Puttaswamaiah, S., 2007. *The conditions for effective mechanisms of compensation and rewards for environmental services.* ICRAF Working Paper no. 38, Nairobi, Kenya: World Agroforestry Centre.

Sydee, J. and Beder, S., 2006. The right way to go? Earth sanctuaries and market-based conservation. *Capitalism Nature Socialism,* 17(1), 83-98.

Tatenhove, J.v., Arts, B. and Leroy, P., 2000. *Political modernisation and the environment: the renewal of environmental policy arrangements. Environment and policy, volume 24.* Dordrecht, the Netherlands: Kluwer Academic Publishers.

Uchida, E., Xu, J.T. and Rozelle, S., 2005. Grain for green: cost-effectiveness and sustainability of China's conservation set-aside program. *Land Economics,* 81 (2), 247-264.

Van Gossum, P., Arts, B., De Wulf, R. and Verheyen, K., 2011. An institutional evaluation of sustainable forest management in Flanders. *Land Use Policy,* 28 (1), 110-123.

Wang, G., Innes, J.L., Wu, S.W. and Dai, S., 2010. Towards a new paradigm: the development of China's forestry in the 21st Century. *International Forestry Review,* 10 (4), 619-631.

Wang, J. and Huang, S., 2008. Enlightenment and progress on the researches of forests ecosystem restoration in the upper reaches of Lijiang River. *Journal of Northwest Forestry University,* 23 (5), 173-177.

Wang, X., 1997. Mutual empowerment of state and peasantry: grassroots democracy in rural China. *World Development,* 25 (9), 1431-1442.

Wang, X., 2008. Analysis on cause and characteristics of soil erosion in Liaoning Province. *China Science and Technology Information* (12), 24-25.

Wang, Z., 2005. Review and reflection on the development of fast-growing and high-yield plantation in Guangxi. *Green China* (12), 26-28.

Wunder, S., 2005. *Payments for environmental services: some nuts and bolts.* CIFOR Occasional Paper No. 42, Jakarta, Indonesia: Center for International Forestry Research.

Wunder, S., 2007. The efficiency of payments for environmental services in tropical conservation. *Conservation Biology*, 21 (1), 48-58.

Wunder, S., Engel, S. and Pagiola, S., 2008. Taking stock: a comparative analysis of payments for environmental services programs in developed and developing countries. *Ecological Economics*, 65 (4), 834-852.

Wunder, S., The, B.D. and Ibarra, E., 2005. *Payment is good, control is better. why payments for forest environmental services in Vietnam have so far remained incipient.* Bogor, Indonesia: CIFOR.

Xu, J. and Melick, D.R., 2007. Rethinking the effectiveness of public protected areas in Southwestern China. *Conservation Biology*, 21 (2), 318-328.

Xu, J., Tao, R. and Amacher, G.S., 2003. *An empirical analysis of China's state-owned forests*: CCAP Working Paper 03-E20.

Xu, J., White, A. and Lele, U., 2010. *China's tenure reforms: impacts and implications for choice, conservation, and climate change.* Washington, DC, USA: Rights and Resources Initiative.

Xu, J., Yin, R., Li, Z. and Liu, C., 2006a. China's ecological rehabilitation: unprecedented efforts, dramatic impacts, and requisite policies. *Ecological Economics*, 57 (4), 595-607.

Xu, X., Ma, T. and Liu, J., 2006b. Study on forest property rights system reform in collective forest area of south China. *Scientia Silvae Sinicae*, 42 (8), 121-129.

Xu, Z., Bennett, M.T., Tao, R. and Xu, J., 2004. China's sloping land conversion program four years on: current situation and pending issues. *International Forestry Review*, 6 (3-4), 317-326.

Yang, Y., 2005. Investigation of the change of ecological environment in original and areas by RS technology. *Journal of Guilin Institute of Technology* (1), 36-41.

Yeh, E., 2009. Greening western China: a critical view. *Geoforum*, 40 (5), 884-894.

Yin, R.K., 1994. *Case study research: design and methods.* London, UK: Sage Publications.

Zbinden, S. and Lee, D.R., 2005. Paying for environmental services: an analysis of participation in Costa Rica's PSA program. *World Development*, 33 (2), 255-272.

Zeng, C., 2003. *Report on the status quo of Fujian Province's ecological environment.* Beijing, China: China Environmental Science Publishing.

Zhang, B., Li, W. and Xie, G., 2010a. Ecosystem services research in China: progress and perspective. *Ecological Economics*, 69 (7), 1389-1395.

Zhang, H., Zhimu, D., Zhang, X., Zhang, J., Wang, W. and Cao, Y., 2005. Comparative analysis of management techniques and administrative policies of public welfare forests in Japan and Liaoning. *Journal of Liaoning Forestry Science and Technology* (5), 35-40.

Zhang, L., 2002. *Ecologizing industrialization in Chinese small towns.* PhD Thesis, Wageningen, the Netherlands: Wageningen University, Environmental Policy Group.

Zhang, L. and Wang, H., 2000. *Practical handbook of China's forestry laws.* Beijing, China: China Forestry Publishing House.

Zhang, L., Mol, A., He, G. and Lu, Y., 2010b. An implementation assessment of China's environmental information disclosure decree. *Journal of Environmental Sciences*, 22 (10), 1649-1656.

Zhang, L., Mol, A.P.J. and Sonnenfeld, D.A., 2007. The interpretation of ecological modernisation in China. *Environmental Politics*, 16 (4), 659-668.

Zhang, L., Tu, Q. and Mol, A.P.J., 2008. Payment for environmental services: the sloping land conversion program in Ningxia autonomous region of China. *China and World Economy*, 16 (2), 66-81.

Zhang, Y., 2005. Multiple-use forestry vs. forestland-use specialization revisited. *Forest Policy and Economics*, 7 (2), 143-156.

Zhao, S., 2011. *Facilitating the second run of NFPP – speech in a national conference for starting the second run of NFPP.* Beijing, China: State Forestry Administration.

Zheng, H. and Zhang, L., 2006. *Chinese practices of ecological compensation and payments for ecological and environmental services and its policies in river basins.* Washington, DC, USA: World Bank.

Zhong, L., 2007. *Governing urban water flows in China.* Wageningen, the Netherlands: Wageningen University.

Zhou, S., 2001. *Chongman Xiwang de Shinian–Xinshiqi Zhongguo Linye Kuayueshi Fazhan Guihua (Ten Years with Prospect – China forestry leap development planning in a new era).* Beijing, China: China Forestry Publishing House.

Zhou, S., 2002. *Historical transitions in China's forestry.* Beijing, China: China Forestry Publishing House.

Zhu, C., Taylor, R. and Feng, G., 2004. *China's wood market, trade and the environment.* Monmouth Junction, NJ, USA: Science Press USA Inc.

Zong, Y. and Chen, X., 2000. The 1998 flood on the Yangtze, China. *Natural Hazards*, 22 (2), 165-184.

# Appendices

## Appendix A. List of interviewees

### *At the national level*

Mrs. Lei Zhang, Director, Department of Rural Forestry Reform and Development, State Forestry Administration

Mr. Guangping Miao, Division Director, Department of Rural Forestry Reform and Development, State Forestry Administration

Mrs. Xiaowen Tang, Deputy Director, Department of Planning and Financial Management, State Forestry Administration

Mr. Huanliang Wang, Deputy Director, China National Forestry Economics and Development Research Centre

Mrs. Guangcui Dai, Deputy Director, China National Forestry Economics and Development Research Centre

Mrs. Yuehua Wang, Researcher, China National Forestry Economics and Development Research Centre

### *The case of Fujian Province*

Mr. Wenchun Cai, Director of Forest Resources Management Station, Fujian Provincial Forestry Department

Mr. Zhuting Chen, Deputy Director of Forest Resources Management Station, Fujian Provincial Forestry Department

Mr. Ligeng Lin, Fujian Provincial Forestry Department

Mr. Xingzhong Xue, Division of Forest Property Rights Management, Fujian Provincial Forestry Department

Mr. Chaoxian Chen, Director, Yongtai County Forestry Bureau

Mr. Zhimian Lin, Director of Forest Management Office, Yongtai County Forestry Bureau

Mr. Hongyan Liang, Chief Engineer, Yongtai County Forestry Bureau

Mr. Liu, Deputy Director, Sanming Municipal Forestry Bureau

Mr. Huang, Director, Shaxian County Forestry Bureau

Mr. Luo, Secretary of CCP Longkeng Village Committee, Shaxian County, Fujian Province

Mr. Yu, Village Head, Longkeng Village, Shaxian County, Fujian Province

### *The case of Guangxi Zhuang Autonomous Region*

Mr. Yejin He, Forestry Fund Management Station, Forestry Bureau of Guangxi Zhuang Autonomous Region

Mr. Li, Forest Tenure Reform Office, Forestry Bureau of Guangxi Zhuang Autonomous Region

Mr. Xiukui Jiang, Forestry Bureau of Guangxi Zhuang Autonomous Region

Mrs. Li Su, Forestry Bureau of Guangxi Zhuang Autonomous Region

Mr. Xiaosan Luo, Forestry Bureau of Guangxi Zhuang Autonomous Region

Mr. Tao, Chief Engineer, Forest Protection Station, Huangmian State-Owned Forest Farm

Mr. Jiansheng Jiang, Director, Yongfu County Forestry Bureau

Mr. Yonggui Liao, Director, Pinglin State-Owned Forest Farm

Mr. Youguang Tan, Director, Dayuan State-Owned Forest Farm

Mr. Yongxiang Mo, Forestry Station, Guilin Municipal Forestry Bureau

Mr. Zengfei Pan, Deputy Director of Guilin Municipal Forestry Bureau

Mr. Li, Director of Linchuan County Forestry Bureau

Mr. Xiaojie Zhao, Linchuan County Forestry Bureau

### *The case of Liaoning Province*

Mr. Wenming Wang, Director of Ecological Benefit Forest Management Office, Liaoning Provincial Forestry Department

Mr. Zhang, Director of Forest Tenure Reform Office, Liaoning Provincial Forestry Department

Mr. Yuku Han, Forest Tenure Reform Office, Liaoning Provincial Forestry Department

Mr. Wang, Planning and Financial Management Office, Liaoning Provincial Forestry Department

Mr. Liu, Deputy Director, Benxi Municipality Forestry Bureau, Liaoning Province

Mr. Cai, Director of Benxi County Forestry Bureau

Mr. Liu, Section chief of Ecological Benefit Forest Management Office, Benxi County Forestry Bureau

Mr. Jiawu Yu, Village Head, Yanghugou Village, Dongyingfang Township, Benxi County

Mr. Hu, Treasurer, Taiyang Village, Dongyingfang Township, Benxi County

Mr. Li, Deputy Director, Fushun Municipality Forestry Bureau, Liaoning Province

Mr. Jianwen Sun, Director, Xinbin County Forestry Bureau, Liaoning Province

Mr. Yongjun Cai, Village Head, Beiwangqing Village, Beisiping Township, Xinbin County

Mr. Zhenquan Yu, Village Head, Luoquan Village, Yushu Township, Xinbin County

## Appendix B. Interview guide for local officers in forest management and protection

Name:
Department and position:
Date:

### *Introduction:*

This interview aims to understand the implementation of all kinds of public payments for forest ecological services in the cases sites. The respondents for the interview include local officers in local forestry bureaus and the heads of towns and villages within the case sites. The interview will include demarcation, management, and various public subsidies for ecological benefit forest, relative taxation and fee and preferential policies for forestry production, and contracts for forest ecosystem services.

1. How many public payment policies are carried out in the case sites (payments for national public benefit forest and regional public benefit forest)?
2. What specific service should the forest provide for the payment (preventing soil erosion, habitats for wild animals and plants and so on) or there is only a general objective without specification?
3. When were the forests demarcated into the public benefit forest zone? Which specific governmental department were in charge of demarcation? How was the demarcation implemented? What factors were taken into account during the demarcation (ecological importance, forest tenure, age and quality of forests)? Which one is the most important for the demarcation?
4. Are the agreement of village collectives and farmer households required when including their forests into the public benefit forest system? Do the village collectives and farmer households have right to refuse the demarcation? After the demarcation, is it possible for them to quit the system?
5. Any kind of disagreement or conflict arose during the demarcation?
6. How is the contract for managing public benefit forest signed? How many stakeholders are included in the making the clauses of the contract?
7. How is the public benefit forest managed? What is the responsibility and tasks for the administrative line of forestry? What is the responsibility and tasks for the village collectives and farmer households?
8. Generally, when does the payment reach to the county level and to the farmer household level? And how much of the payment gets to different levels? In the end, which proportion of the payment can farmer households obtain?
9. Which department takes charge of inspecting and monitoring the effect of public benefit forest management? How often? What is the procedure like? What kinds of criteria are used in the inspecting and monitoring? Does the result have influence on the payment that the collectives or farmer households receive next time? If the quality of public benefit forest goes down or deforestation happens, what kind of measure will be taken?

10. Besides inspecting and monitoring, has any other evaluation been taken for assessing forest ecosystem services? If yes, how is the procedure and what kinds of indicators have been included? If not, what are the reason and the major difficulties?

## Appendix C. Questionnaire for farmer households involved in the payment schemes for forest ecosystem services

Location:        Province/Autonomous Region:       County:       Township:       Village:
Respondent No.:
Sex:

### *Part 1: Background on farmer respondents*

First, I would like to ask some questions about your family and yourself.
1.  How old are you?
2.  Which ethnic group do you belong to?
3.  Which education level have you reached?
    (1) university; (2) senior high; (3) technical school; (4) junior high; (5) elementary;
    (6) no school experience.
4.  How many members in your family?
5.  How many children in your family (younger than 18 year old)?

### *Part 2: Income change before and after the schemes*

Next, please recall the income of your family before the payment schemes.
1.  How much was the total income of your family in the year just before the schemes?
2.  Which is major income source for your family before the schemes?
    (1) agriculture; (2) forestry; (3) animal husbandry; (4) off-farm work; (5) small business; (6) other
3.  How many (mu: 1/15 hectare) croplands did your family have in that year?
4.  How much did your family earn from agriculture in that year?
5.  How much did your family earn from animal husbandry in that year?
6.  How many family members went outside for off-farm work?
7.  How much did your family earn from off-farm work in that year?
8.  How many forestlands did your family have in that year?
9.  How much did your family earn from timber-harvesting in that year?
10. How much did your family earn from raising fungi and planting Chinese herbal medicine under forests in that year?
11. Did your family get income from forest-related tourism around or in the public benefit forest? If yes, what kind of income and how much?

Please recall the income of your family last year (2007 in Fujianjian; 2008 in Guangxi; 2009 in Liaoning)
1.  How much is the total income of your family last year?
2.  Which is major income source for your family last year?
    (1) agriculture; (2) forestry; (3) animal husbandry; (4) off-farm work; (5) small business; (6) other

3.  How many (mu: 1/15 hectare) croplands did your family have last year?
4.  How much did your family get from agriculture last year?
5.  How much did your family get from animal husbandry last year?
6.  How many family members went outside for off-farm work last year?
7.  How much did your family get from off-farm work last year?
8.  How many forestlands did your family have last year?
9.  How much did your family get from timber-harvesting last year?
10. How much did your family get from raising fungi and planting Chinese herbal medicine under forests in that year?
11. Did your family get income from forest-related tourism? If yes, what kind of income and how much?
12. Forest resources of the farmer household

| No. of forest plot | Area (ha) | Acquisition[1] | Year of acquisition | Classification[2] | Volume (m³/ha) | Tree species | Forest type[3] | Ages |
|---|---|---|---|---|---|---|---|---|
| 1 | | | | | | | | |
| 2 | | | | | | | | |
| 3 | | | | | | | | |

[1] Acquisition: (1) *ziliushan* (private plots); (2) contracted before tenure reform; (3) contracted after tenure reform; (4) other.

[2] Forest classification: (1) public benefit forest; (2) commercial forest.

[3] Forest type: (1) natural forest; (2) plantation.

13. Payment for your family from the schemes:

| Year | 2001 | 2002 | 2003 | 2004 | 2005 | 2006 | 2007 | 2008 | 2009 |
|---|---|---|---|---|---|---|---|---|---|
| Public benefit forest (ha) | | | | | | | | | |
| Payment (Yuan) | | | | | | | | | |

## Part 3: Participation of the farmer household in payment schemes

1.  Do you know the central fiscal payment policies for public benefit forest (including demarcation, management, payment distribution, etc.)?
    (1) very well; (2) some parts; (3) a little bit; (4) not heard them yet
2.  Do you know the policies of the provincial payment schemes?
    (1) very well; (2) some parts; (3) a little bit; (4) not heard them yet
3.  Do you know payment policies for public benefit forest in your county?
    (1) very well; (2) some parts; (3) a little bit; (4) not heard them yet
4.  How did you get the information about the payment policies for public benefit forest?
    (1) village bulletin; (2) village group meetings; (3) local media (newspaper and TV);
    (4) communication within neighborhood; (5) other channels (please indicate)
5.  Has your family been consulted in the process of the demarcation of public benefit forest?
    (1) yes; (2) no

If yes, which form of consultation has been taken for your family?
(1) Village representative meeting
(2) Village member meeting
(3) Bulletin for opinion
(4) Direct negotiation between the village committee and farmer households
(5) Other forms

6. At that time, did you agree to the demarcation of public benefit forest?
(1) strongly agree; (2) basically agree; (3) disagree
If you disagreed, what was the reason?
(1) High timber revenue cannot be offset by the payment
(2) Need to repay the afforestation loan
(3) Need to pay the contract charge for the forestland
(4) No timber for house building or daily fuel
(5) Other (please indicate)

7. If only taking environmental protection into account, do you think the demarcation of public benefit forest is appropriate?
(1) appropriate, most of ecologically valuable forest has been included for protection (water source sites, animal habitats, etc.)
(2) inappropriate
If it is inappropriate, what is your opinion?
(1) too much; (2) too less; (3) other reasons.

8. Do you think the arrangement of public benefit forest protection is fair (whether benefits and responsibilities relating to the schemes have been distributed evenly among the members)?
(1) very fair; (2) fair; (3) basically fair; (4) not fair; (5) don't know

9. Which method do you think is the best way to distribute public benefit forest to each household?
(1) Decided by high-level governments
(2) Evenly distributed by high-level governments
(3) Decided by village committees
(4) Evenly distributed by village committees
(5) Decided by households involved in the public benefit forest planning
(6) Decided by all households in the village
(7) Other (please indicate)

10. What impact did the protection of public benefit forest have on the revenue of your family?
(1) increase; (2) decrease; (3) no impact; (4) don't know

11. What impact did the protection of public benefit forest have on the livelihood of your family?
(1) Lack of wood for daily fuel;
(2) No forestland for livestock browsing;
(3) Other (please indicate)

12. Who do you think should be in charge of funding the payment of public benefit forest?
(1) the central government; (2) local governments; (3) enterprises and residents along downstream; (4) other (please indicate)

13. According to your loss from schemes, do you think whether the payment standard is high or low against your expectation?

(1) high; (2) appropriate; (3) low

14. Who do you think would be the most suitable organization for the distribution of the payment?
    (1) county forestry bureau; (2) county government; (3) township government (4) village committee; (5) other

15. Have you signed the contract on protection of public benefit forest?
    (1) yes; (2) no

16. Have you participated in signing the payment contract for public benefit forest owned by village collectives?
    (1) yes; (2) no

17. Does your family take the task of protecting and managing public benefit forest?
    (1) yes; (2) no
    If yes, how is the workload of task for you?
    (1) too much; (2) too less; (3) appropriate
    If your family takes the task, do you think the wage is enough?
    (1) too much; (2) too less; (3) appropriate

18. Have the village heads consulted you about the decision on personnel for protecting and managing public benefit forest?
    (1) yes; (2) no

19. Are you satisfied with the decision on the personnel for protecting and managing public benefit forest?
    (1) yes; (2) basically yes; (3) no

20. Has any member of your family participated in any training for protecting and managing public benefit forest?
    (1) yes; (2) no

21. Who do you think can give an efficient inspection and monitoring on the protection of public benefit forest?
    (1) provincial forestry department; (2) county forestry bureau; (3) township forestry stations; (4) village committees; (5) self-monitoring by villagers

22. Have you reported any problem in the protection and management of public benefit forest?
    (1) yes; (2) no
    If yes, which organization have you reported to?
    (1) provincial forestry department; (2) county forestry bureau; (3) township forestry station; (4) village committee

23. Have you been consulted when forestry departments from the province and the counties inspected the performance of protection and management of public benefit forest?
    (1) yes; (2) no

24. Do you think whether local soil erosion was severe before the payment policies (the villagers maybe know what is severe soil erosion)?
    (1) universally severe; (2) partly severe; (3) no apparent erosion

25. If the soil erosion was severe, do you think the soil erosion has been reduced after the implementation of the payment policies?
    (1) significantly reduced; (2) slightly reduced; (3) no change

### Part 4: Collective forest tenure reform

1. Do you know your village's arrangement of collective forest tenure reform?
   (1) very well; (2) a little bit; (3) not heard it yet
2. Who decided the arrangement for your village?
   (1) the village committee; (2) village heads; (3) voting by a village all-member meeting;
   (4) voting by a village representative meeting
3. Are you satisfied with the arrangement of collective forest tenure reform?
   (1) yes; (2) no
4. In your opinion, were the property rights of forest stable in your village?
   (1) always stable without adjustment on boundaries; (2) relatively stable with some adjustment;
   (3) unstable with frequent redistribution
5. How were the disputes on forest property rights in your village before the tenure reform?
   (1) basically no disputes; (2) disputes on some parts of forests; (3) widely disputes on forests
6. How are the boundaries of the forest plots of your family after the reform?
   (1) clear without disputes; (2) clear but having disputes; (3) unclear without disputes; (4) other
7. Have you received formal certificates of forest property rights from the local forestry bureau?
   (1) yes; (2) no
8. If you have certificates of public benefit forest property rights, how many certificates do you
   have and how much area of public benefit forest are indicated on them?
9. Do you need to pay use fee for contracting public benefit forest and how to pay it?
   (1) no; (2) annual use fee required; (3) timber profits split in proportion; (4) other
   If annual use fee is required, how much for one ha each year?
   If timber profits split in proportion, which proportion for the village?
10. Do you need to pay a deposit for forest regeneration?
    (1) yes; (2) no
    If yes, how much for one ha?
11. After the collective forest tenure reform, have you planted trees on your public benefit forest?
    (1) yes; (2) no
    If yes, how much plantations have you developed?
    How much was the total cost? (including seedling, tillage, planting, chemical fertilizer, etc.)
    Have you received subsidy from the governments for developing plantations?
    (1) yes; (2) no
    If yes, how much was the subsidy?
12. After the collective forest tenure reform, have you harvested timber from the public benefit
    forest?
    (1) yes; (2) no
    If yes, how many cubic meters have you harvested and how much was the income from it?
13. Do you think it is difficult to apply the permit for logging public benefit forest?
    (1) yes; (2) no
    If yes, what are the major difficulties?
    (1) complex procedure for application; (2) too long time for application;
    (3) impossible for application; (4) other

14. Do you think the collect forest tenure reform prompted clarification of forest property rights?
    (1) significantly prompted; (2) improved a little; (3) no use
15. You will increase investment on forestry after the collective forest tenure reform?
    (1) increase; (2) stay the same; (3) decrease
16. Do you think that the public benefit forest should be distributed to individual farmer household?
    (1) yes; (2) no
    If yes, what is the reason?
    (1) Help to channel the payment to the level of farmer households
    (2) Improve public benefit forest management
    (3) Create opportunities for developing agro-forestry in public benefit forest
    (4) Other reason (please indicate)
    If no, what is the reason?
    (1) Impede fire and pest control for public benefit forest
    (2) Induce illegal logging
    (3) Not much marginal profits, not necessary
    (4) Other reason (please indicate)

## Part 5: Willingness of local farmers to pay for protecting public benefit forest

1.  How much do you prefer to receive for your current public benefit forest each year?
2.  For the public benefit forest managed by joint family households, how much do you prefer to receive each year? And how much share do you want to get for your family from the payment?
3.  If an auction is introduced to the payment schemes, in which villages and farmer households can voluntarily negotiate with the governments about the area of public benefit forest and the payment and the governments will choose the forests with high ecological importance and low bidding price as public benefit forest. Do you like this kind of payment schemes?
    (1) yes; (2) no; (3) I don't know
4.  Would you like to adjust the area of the public benefit forest owned by your family?
    (1) I hope to change some commercial forests into public benefit forest
    (2) I hope to change some public benefit forests into commercial forest
    (3) I do not want any change

## Part 6: Participation in other ecological conservation program

1.  Have your family participated in the Conversion of Cropland into Forest and Grassland Programs?
    (1) yes; (2) no
    If yes, how much your cropland has been conversed? How much do you receive as subsidy each year?
2.  Have your family participated in local natural forest protection programs?
    (1) yes; (2) no
    If yes, how much area of natural forests have you managed and protected? How much do you receive as subsidy each year?

3. Which program do you think achieved the best environmental effectiveness?
   (1) Payment schemes for public benefit forest
   (2) Conversion of Cropland into Forest and Grassland
   (3) Local natural forest protection programs
4. Which program do you think contribute most to your family income?
   (1) Payment schemes for public benefit forest
   (2) Conversion of Cropland into Forest and Grassland
   (3) Local natural forest protection programs
   (4) None

# Appendix D. Survey questionnaires on the cost and benefit of public benefit forest management

These questionnaires were used to collect cost and benefit data on public benefit forest of state-owned forest farms in Guangxi Zhuang Autonomous Region.

Respondent:
Department and position:

## Part 1: Situation of forest resources BEFORE the payment schemes

(Note: these are the data on the public benefit forest before the payment schemes started. It is needed to be carefully recalled, especially resource structure, stock volume and forest ages)

| Plot No. | Area (ha) | Main usage[1] | Volume (m³/ha) | Tree species | Generation[2] | Age | Slope | Soil[3] | Distance to road[4] |
|---|---|---|---|---|---|---|---|---|---|
| 1 | | | | | | | | | |
| 2 | | | | | | | | | |
| 3 | | | | | | | | | |
| | | | | | | | | | |

[1] Main usage: (1) timber forest; (2) protection forest; (3) economic forest; (4) fuel forest; (5) special use forest; (6) bamboo forest; (7) other (please indicate).

[2] Generation: (1) natural forest; (2) plantation.

[3] Soil condition: (1) very fertile; (2) relatively fertile; (3) not too bad; (4) relatively sterile; (5) very sterile.

[4] Distance to road indicates the distance of the forest plots from roads available for vehicles, in km.

## Part 2: Cost and benefit of forest management BEFORE the payment scheme

### 1. Afforestation

| Plot No. | Afforestation year | Seedling (Yuan) | Tillage[1] | | | | Planting | | | |
|---|---|---|---|---|---|---|---|---|---|---|
| | | | own labor input[2] (work days) | hiring[3] (work days) | labor price (Yuan) | employment cost (Yuan) | own labor input[2] (work days) | hiring[3] (work days) | labor price (Yuan) | employment cost (Yuan) |
| 1 | | | | | | | | | | |
| | | | | | | | | | | |
| 2 | | | | | | | | | | |
| | | | | | | | | | | |
| 3 | | | | | | | | | | |
| | | | | | | | | | | |

[1] Tillage indicates soil preparation for plantations, including brush cutting, tilling, digging, etc.

[2] Own labor input indicates the labor of workers of the state-owned forest farm.

[3] Hiring: the farm hires labor outside for afforestation.

### 2. Regular management cost

| Plot No. | Management and protection cost | | | | | Other management cost | |
|---|---|---|---|---|---|---|---|
| | own labor input (person) | hiring outside (person) | wage (Yuan per capita per year) | total wage (Yuan per year) | management contract term (year) | years | average cost each year (Yuan) |
| 1 | | | | | | | |
| 2 | | | | | | | |
| 3 | | | | | | | |

## 3. Cost and benefit of selective logging

| Plot No. | Year | Selective logging volume (cubic meter) | Logging labor input | | | | Transport cost (Yuan) | Tax and fee[1] (Yuan) | | | | Other logging cost[2] (Yuan) | Timber sale revenue (Yuan) | |
|---|---|---|---|---|---|---|---|---|---|---|---|---|---|---|
| | | | own labor input (work days) | hiring (work days) | labor price (Yuan) | employment cost (Yuan) | | tax | forest cultivation fee | other fee | mountain rent | | price | total |
| 1 | | | | | | | | | | | | | | |
| 2 | | | | | | | | | | | | | | |
| 3 | | | | | | | | | | | | | | |
| | | | | | | | | | | | | | | |

[1] Tax and fee includes agricultural special tax, forest cultivation fee and other tax and fee in the sales of timber.
[2] Other cost includes logging design cost and other cost related to logging.

## 4. Cost and benefit of final logging

| Plot No. | Year | Final logging volume (cubic meter) | Final logging labor input | | | | Transport cost (Yuan) | Tax and fee[1] (Yuan) | | | | Other logging cost[2] (Yuan) | Timber sale revenue (Yuan) | |
|---|---|---|---|---|---|---|---|---|---|---|---|---|---|---|
| | | | Own labor input (work days) | Hiring (work days) | Labor price (Yuan) | Employment cost (Yuan) | | Tax | Forest cultivation fee | Other fee | Mountain rent | | price | Total |
| 1 | | | | | | | | | | | | | | |
| 2 | | | | | | | | | | | | | | |
| 3 | | | | | | | | | | | | | | |
| | | | | | | | | | | | | | | |

[1] Tax and fee includes agricultural special tax, forest cultivation fee and other tax and fee in the sales of timber.
[2] Other cost includes logging design cost and other cost related to logging.

## Part 3: Situation of forest resources AFTER the payment scheme

1. Forest resources in 2007

| Plot No. | Area (ha) | Main usage[1] | Volume (m³/ha) | Tree species | Generation[2] | Age |
|----------|-----------|---------------|----------------|--------------|---------------|-----|
| 1 | | | | | | |
| 2 | | | | | | |
| 3 | | | | | | |
| | | | | | | |

[1] Main usage: (1) timber forest; (2) protection forest; (3) economic forest; (4) fuel forest; (5) special use forest; (6) bamboo forest; (7) other (please indicate).
[2] Generation: (1) natural forest; (2) plantation.

2. Forest resources in 2008

| Plot No. | Area (ha) | Main usage[1] | Volume (m³/ha) | Tree species | Generation[2] | Age |
|----------|-----------|---------------|----------------|--------------|---------------|-----|
| 1 | | | | | | |
| 2 | | | | | | |
| 3 | | | | | | |
| | | | | | | |

[1] Main usage: (1) timber forest; (2) protection forest; (3) economic forest; (4) fuel forest; (5) special use forest; (6) bamboo forest; (7) other (please indicate).
[2] Generation: (1) natural forest; (2) plantation.

## Part 4: Cost and benefit of forest management AFTER the payment scheme

1. Afforestation

| Plot No. | Afforestation year | Seedling (Yuan) | Tillage[1] | | | | Planting | | | |
|---|---|---|---|---|---|---|---|---|---|---|
| | | | own labor input[2] (work days) | hiring[3] (work days) | labor price (Yuan) | employment cost (Yuan) | own labor input[2] (work days) | hiring[3] (work days) | labor price (Yuan) | employment cost (Yuan) |
| 1 | | | | | | | | | | |
| | | | | | | | | | | |
| | | | | | | | | | | |
| 2 | | | | | | | | | | |
| | | | | | | | | | | |
| | | | | | | | | | | |
| 3 | | | | | | | | | | |
| | | | | | | | | | | |
| | | | | | | | | | | |

[1] Tillage indicates soil preparation for plantations, including brush cutting, tilling, digging, etc.

[2] Own labor input indicates the labor of workers of the state-owned forest farm.

[3] Hiring: the farm hires labor outside for afforestation.

2. Regular management cost

| Plot No. | Management and protection cost | | | | | Other management cost | |
|---|---|---|---|---|---|---|---|
| | own labor input (person) | hiring outside (person) | wage (Yuan per capita per year) | total wage (Yuan per year) | management contract term (year) | years | average cost each year (Yuan) |
| 1 | | | | | | | |
| 2 | | | | | | | |
| 3 | | | | | | | |

### 3. Cost and benefit of selective logging

| Plot No. | Year | Selective logging volume (cubic meter) | Logging labor input | | | | Transport cost (Yuan) | Tax and fee[1] (Yuan) | | | | Other logging cost[2] (Yuan) | Timber sale revenue (Yuan) | |
|---|---|---|---|---|---|---|---|---|---|---|---|---|---|---|
| | | | own labor input (work days) | hiring (work days) | labor price (Yuan) | employment cost (Yuan) | | tax | forest cultivation fee | other fee | mountain rent | | price | total |
| 1 | | | | | | | | | | | | | | |
| 2 | | | | | | | | | | | | | | |
| 3 | | | | | | | | | | | | | | |
| | | | | | | | | | | | | | | |

[1] Tax and fee includes agricultural special tax, forest cultivation fee and other tax and fee in the sales of timber.

[2] Other cost includes logging design cost and other cost related to logging.

### 4. Cost and benefit of final logging

| Plot No. | Year | Final logging volume (cubic meter) | Final logging labor input | | | | Transport cost (Yuan) | Tax and fee[1] (Yuan) | | | | Other logging cost[2] (Yuan) | Timber sale revenue (Yuan) | |
|---|---|---|---|---|---|---|---|---|---|---|---|---|---|---|
| | | | Own labor input (work days) | Hiring (work days) | Labor price (Yuan) | Employment cost (Yuan) | | Tax | Forest cultivation fee | Other fee | Mountain rent | | price | Total |
| 1 | | | | | | | | | | | | | | |
| 2 | | | | | | | | | | | | | | |
| 3 | | | | | | | | | | | | | | |
| | | | | | | | | | | | | | | |

[1] Tax and fee includes agricultural special tax, forest cultivation fee and other tax and fee in the sales of timber.

[2] Other cost includes logging design cost and other cost related to logging.

## Appendix E. Questionnaire on the quality of the public benefit forest

(Please fill in the codes or numbers)

| Evaluative indicators | Evaluative standards | The year just before the payment schemes | 2007 | 2008 |
|---|---|---|---|---|
| Biodiversity | (1) broadleaf dominated forest | | | |
| | (2) mixed forest with more than 30% broadleaf trees | | | |
| | (3) conifer dominated forest or vegetation with brush &grass | | | |
| Canopy density | The proportion of an area in the ground that is covered by the crown of trees and is expressed in percentage. | | | |
| Phytocenosis structure | (1) with more than two-tier trees, shrub and grass | | | |
| | (2) with trees, shrub and grass | | | |
| | (3) sparse shrub or grass | | | |
| Vegetation height | meter | | | |
| Vegetation coverage | the proportion of an area in the ground that is covered by all vegetation (trees, brush, grass) and is expressed in percentage. | | | |
| Depth of litter layer | centimeter | | | |

# Summary

Forest ecosystems in China are valuable for protecting ecologically fragile areas and for providing ecological services for people. However, China's forest resources are under pressure from its fast-growing economy and rapid industrialization. While forest disappears, ecological crises become frequent and devastating. Traditional forest management characterized by command and control measures seems to lack power to alleviate the tension between ecological needs and economic development. More importantly, China's forested regions, which are usually less developed, face greater pressure to reduce poverty and improve local livelihood. In addition, the collective forest tenure reform, which was launched recently, is changing the arrangement of forest property rights in rural areas. All of these lead to a mixed picture of China's forest management. Under such circumstances, PES is introduced as a promising solution to the conflict between ecological conservation and economic development. An array of ecological conservation programs in accordance with the principle of PES were implemented throughout the country. Among these ecological conservation programs, payment schemes for forest ecosystem services demonstrated strong policy intervention in forest use practice and financial stability. However, the performance of the payment schemes has not been closely scrutinized.

This research aims to evaluate the payment schemes for forest ecosystem services in China after they have been implemented for more than a decade. Rather than taking for granted the government's claim on the success of the payment schemes in protecting forests and providing ecological services, the research tries to examine PES schemes with respect to ecological effectiveness, economic and livelihood impacts, participation of local people, and the interlinkage with forest tenure reform, by analyzing the performance of the payment schemes in different cases. Based on the objectives mentioned above, three research questions have been defined:
- What have been the ecological and socio-economic effects of forest PES schemes in China?
- To what extent and how have state and non-state actors (including farmers) participated in the design, implementation and evaluation of forest PES schemes in China?
- How has forest tenure reform influenced the functioning and outcome of forest PES schemes in China?

In order to answer these questions, an evaluative framework has been developed. The evaluation framework focuses on three aspects of payment schemes: policies, practices and performance. The analysis of policies focuses on the relationship between the institutional setting and payment schemes. The implementation of payment schemes (via the institutional setting) affects forest use practices, which is the core element of the evaluation. Forest use practices are affected by payment schemes and this in turn influences the formulation and implementation of subsequent payment schemes. This research aims to evaluate the dynamics of this process from policies to practices. Three dimensions were selected to explore and evaluate the process: environmental effectiveness, livelihood impacts and participation.

Three case provinces (Fujian, Guangxi and Liaoning) have been selected to sufficiently represent the geographical variation of program areas, in terms of forest coverage, economic and social development. Interviews with local officials and surveys at farmer household level have been employed to collect data for the policy evaluation.

Following the implementation of the Forest Ecological Benefit Compensation Fund Program funded by the central government, local governments established their own local payment schemes. Three institutional factors have prompted the development of payment schemes. First, classification-based forest management offers an institutional setting for payment schemes by stipulating formal rules for classification, protection, and management of public benefit forests. Second, collective forest tenure reform, which reshaped the structure of forest ownership in rural areas, widened the gap between profits from commercial forests and those from public benefit forests and thereby led to the creation of new local payment schemes to fill the gap. Finally, the political willingness of local governments for PES played an important role in influencing the political agenda regarding the development of local payment schemes.

The assessment showed that the effects of the payment schemes on environmental services are most likely positive. After the forest tenure reform settles down the forest ownership, forest owners even become more active in participating in forest protection and restoration. However, the age structure and tree species composition is still in a poor condition and current payment schemes did not provide an efficient mechanism to deal with it.

Under the existing payment standard, the schemes tend to have negative impacts on local livelihood. Especially in poverty-stricken areas such as Guangxi, the implementation of payment schemes encountered difficulties. In such regions, local governments did not have sufficient financial support for developing new local payment schemes and their low administrative capacity and efficiency cannot ensure effective monitoring and smooth implementation of the payment schemes. Negative impacts on the local economy and farmers' income increased the complaints and rejection from local farmers. However, alternative livelihood opportunities, including off-farm employment, agro-forestry, and ecotourism, can partly offset these direct income impacts. At the same time, local forest use practices have shifted away from traditional timber extraction and transformed into a more sustainable use of forest resources. Forest industry policy has a great influence on this process.

The central and provincial or regional governments still play a key role in initiating, designing, implementing, managing and examining payment schemes. Local forest owners have not been involved in the decision making on payment schemes; neither was the participation of local farmer households in the demarcation of the public benefit forests prevalent. They still lack influence in shaping the payment schemes and in initiating public benefit forests. In contrast, farmer households have participated or been included in the management of public benefit forests. There is indication that their involvement in the implementation increased the environmental effectiveness of the payment schemes. During the forest tenure reform, local farmers gained more opportunities for direct participation. The reform also created a legitimate position for farmers to participate in public benefit forest management. This positive change in participation contributed to further improvement of the livelihood of local farmers through both the forest tenure reform and the payment schemes. However, channels for participation of local farmers are still needed to extend to effectively communicate with them, especially in the demarcation of the public benefit forests.

Collective forest tenure reform seems to have improved the environmental performance of payment schemes and have reduced their negative impacts on local livelihood. Distribution of public benefit forests to individual farmer households facilitated the flow of payments to individual farmer households and reduced the risk that the government and village committees divert the payment. Furthermore, the forest tenure reform created new room for developing market-based voluntary PES in China. Clarification of forest ownership provided a helpful institutional setting for establishing a more market-oriented PES mechanism.

Overall, the implementation of payment schemes for forest ecological services in China shows improvements in forest management and protection, as it not only breaks with previous single minded timber extraction but also increasingly employs an incentive-based governance method instead of "command and control". The payment schemes created new forest use practices which couple local livelihood requirements with ecological conservation. Furthermore, local farmers are more and more involved in the implementation of payment schemes, rather than marginalized consumers of decisions regarding their traditional sustenance sources. The research confirmed that the payment schemes in China can be understood as a process of ecological modernization in China's forest sector. However, in the case of payments schemes, the process is colored with typical Chinese characteristics. Compared to ecological modernization inspired PES schemes in western countries, regulatory measures still play an important role in maintaining the Chinese schemes and reigning local forest use. The government plays a major role in steering the direction of ecological modernization processes through its administrative system and its financial support. Market actors play less significant roles in the process than in the PES schemes of other countries.

# About the author

Dan Liang was born on 18 March 1979 in Tianmen, Hubei Province, China. He obtained his Bachelor degree in Environmental Planning and Management from Nankai University in 2000. In the same year, he moved to the Research Center for Eco-Environmental Sciences (RCEES), Chinese Academy of Sciences (CAS) and studied in the Environmental Management and Policy Group. In June 2003, he obtained his MSc degree from CAS. Since then, he has been a researcher of the China National Forestry Economics and Development Research Center (CNFEDRC), engaging in a number of research projects on ecological conservation and forest management. In March 2007, he pursued his Doctoral Degree in Wageningen University, the Netherlands.

Printed in the United States
by Baker & Taylor Publisher Services